"The Book of Physics: Volume 1"

SIMONE MALACRIDA

ANALYTICAL INDEX

Analytical index

Analytical index

Analytical index

Analytical index

Analytical index

INTRODUCTION

This book was born from the need to reconcile, in a single text, all the physical theories studied to date, complete with their theoretical and experimental framework.

There is no doubt that physics, as we understand it today, originated from the introduction of the scientific method, first on a philosophical level, then on an experimental and practical level.

When the scientific method entered the practice of reasoning on which to base assumptions and deductions, there was an enormous leap in quality compared to all previous knowledge.

We can say that all the discoveries and applications that took place in the past with respect to that event are actually the result of semi-empirical approaches and not exactly of science as we understand it today.

That point of no return was such as to determine a historical watershed, in the same way as we are used to considering events of the caliber of the French Revolution, the fall of the Roman Empire or the discovery of America.

Since that time, scientific investigation has had an impressive acceleration, ranging in every field of knowledge and has impressed on society, in terms of applications and daily consequences, a decidedly different imprint than in the past, coming to create those conditions and those prerequisites necessary for the Industrial Revolution, which occurred only less than two centuries after those first scientific stirrings.

A first caesura of this path occurs with the end of the nineteenth century and with the acknowledgment that, in the range of knowledge in all sectors, such contradictions had been reached that the previous theoretical schemes had to be completely revised.

From that period, historically known as the crisis of classical physics, came the two revolutionary theories of the twentieth century which are the basis of contemporary physics, the one we use today to describe Nature and what surrounds us.

In this period of time, which lasted a good two centuries, physics has managed to scientifically explore various disciplines such as mechanics in all its forms (static, dynamic and kinematic), astronomy, the theory of gravitation, optics, the phenomena and oscillatory ones, fluid dynamics, thermodynamics, heat transmission, statistics applied to physics, electric and magnetic phenomena.

1

As can be seen from this small list, the elaboration of theories that predict and explain the experimental results has been so pervasive as to have left nothing unexplored, with the limitations that the equipment of the time could have (it is obvious that it was completely beyond place to think of probing the characteristics of the atom and of the atomic nucleus, not having at one's disposal the suitable material means to detect the essential experimental data).

What has just been described is dealt with in the first part of this book which coincides with the treatment of classical physics.
The second part of the book is inspired by the great revolutions of the early twentieth century, namely quantum physics and special relativity.
They have played such an extraordinary role in the development of physics that it was decided to devote an entire part to them.
The third part of the book deals with the physics of the microcosm, i.e. the physics that develops on a molecular, atomic, nuclear and fundamental particle scale.
We will see how far scientific investigation has gone and what the problems of these developments are today.
The fourth part instead, as a counterpart, deals with the physics of the macrocosm and has the theory of general relativity as its founding stone.
It is everything related to astronomy, astrophysics and cosmology.
Also in this case the recent results of these theories will be tangible.
The fifth and last part has the most difficult task compared to the others.
In fact, if on the one hand the theory of relativity has generated speculations on the macrocosm and quantum physics those on the microcosm, there is numerous evidence of their possible (and desirable) meeting in a single theory.
The last part of the book deals with this peculiar aspect.

The book is divided into chapters, each of which can very well be treated independently of the previous and subsequent ones (in fact, in the literature there are numerous writings relating precisely to each of the chapters exposed).
However, there is a logical correlation in the order of the chapters, a sort of progressive knowledge towards what was previously unknown.
The attentive reader will realize this and will be able to follow this leitmotif which is none other than the re-proposition of the history of physics.
A note should be made about the execution of the exercises.
It is true that in the first part, the one dedicated to classical physics, exercises carried out at the high school level are presented (precisely because in high school one begins to study those specific sectors of physics), but it is equally true that the theoretical formalism is, almost early on, focused on college-level mathematics that assumes knowledge of

advanced mathematical analysis, advanced geometry, and other mathematical disciplines.

What is the point of studying physics?

Let's try to give a brief explanation (entirely personal, of course).

We cannot hide the fact that the interpretation of physical laws, if pushed to the maximum level, can only lead to speculative questions typical of philosophy, especially when dealing with the infinitely large (as in the case of cosmology) or the infinitely small (as in particle physics).

Physical laws, precisely because they have the peculiarity of explaining nature, the universe and everything that surrounds us, must not only be in agreement with the experimental data, but constitute a theoretical model for the simulation of reality itself.

Their structure and interpretation therefore influence the way of describing reality, as already happened with the advent of relativism and indeterminism at the beginning of the twentieth century.

The physical laws are written with a symbolism that is mathematical. The great "strength" of mathematics lies in at least three distinct points.

First of all, thanks to it it is possible to describe reality in scientific terms, that is by foreseeing some results even before having the real experience.

Predicting results also means predicting the uncertainties, errors and statistics that necessarily arise when the ideal of theory is brought into the most extreme practice.

Second, mathematics is a language that has unique properties.

It is artificial, as built by human beings.

There are other artificial languages, such as the Morse alphabet; but the great difference of mathematics is that it is an artificial language that describes nature and its physical, chemical and biological properties.

This makes it superior to any other possible language, as we speak the same language as the Universe and its laws.

At this juncture, each of us can bring in our own ideologies or beliefs, whether secular or religious.

Many thinkers have highlighted how God is a great mathematician and how mathematics is the preferred language to communicate with this superior entity.

The last property of mathematics is that it is a universal language.

In mathematical terms, the Tower of Babel could not exist.

Every human being who has some rudiments of mathematics knows very well what is meant by some specific symbols, while translators and dictionaries are needed to understand each other with written words or oral speeches.

We know very well that language is the basis of all knowledge.

The human being learns, in the first years of life, a series of basic information for the development of intelligence, precisely through language.

The human brain is distinguished precisely by this specific peculiarity of articulating a series of complex languages and this has given us all the well-known advantages over any other species of the animal kingdom.

Language is also one of the presuppositions of philosophical, speculative and scientific knowledge and Gadamer has highlighted this, unequivocally and definitively.

But there is a third property of mathematics which is far more important.

In addition to being an artificial and universal language that describes nature, mathematics is properly *problem solving* , therefore it is concreteness made science, as man has always aimed at solving problems that grip him, just take a look at what has been discussed in this paper about the overcoming of physical theories.

The texture of reality is therefore marked by physical laws that underlie mathematical equations and which, over time, tend to generalize more and more on the wave of new discoveries and inconsistencies of old theories.

Today we are faced with one of those fundamental steps.

On the one hand we know that there are problems of congruence of the two main theories (general relativity and quantum field theory), on the other we have not yet defined a new theoretical canvas that overcomes these obscure points towards a wider knowledge.

As always, it is a constant challenge and, in some way, eternally inherent in human nature.

This characteristic is part of an eternal race towards a better description of what surrounds us and a better understanding of all existing phenomena, in the wake of a derivation from the myth of Ulysses, which embodies man's eternal propensity towards knowledge.

PART ONE: CLASSICAL PHYSICS

1

THE SCIENTIFIC METHOD

Introduction

The beginning of modern physics coincides with the formulation and application of the scientific method, which took place in a systematic way in the early seventeenth century above all by Galileo and with decisive contributions by the philosophers Bacon and Descartes.

This logical and philosophical structure became the basis for the construction of scientific knowledge in the following centuries and for the first mathematical approach through the introduction of analysis by Newton and Leibnitz in the second half of the seventeenth century.

Before Galileo, knowledge had progressed above all through empirical attempts or purely metaphysical reasoning, relying on logical constructs such as the syllogism or the principle of authority. There were therefore no scientists as we understand them today and the closest thing to our concept of science was given by the scholars of natural philosophy.

A forerunner of the scientific method was Leonardo da Vinci who, about a century before Galileo, understood the fundamental importance of real experimentation and mathematical demonstration, without however arriving at the definition of a system and a method.

The vision of Galileo Galilei

Galileo started from some fundamental assumptions, which are still valid today, among which:

1) Nature responds to mathematical criteria
2) To establish the laws of physics it is necessary to carry out experiments
3) Logical hypotheses and mathematical theories must be in agreement with experiments

Therefore Galileo abandoned the empty search for the primary essences and qualities that had characterized knowledge so much before the seventeenth century and set quantitative facts, measurable and verifiable through experiments and expressible through the language of mathematics, as the cornerstone of science.

One of the key points is given by the reproducibility of the experiments: under suitable conditions and hypotheses to be prepared, a certain experience must be able to be repeated in every place giving the same results and therefore confirming (or denying) the mathematical theory formulated to explain this experiment.

In particular cases where it is not possible to carry out a real experiment, Galileo introduces the concept of thought experiment.

By applying the same mathematical and quantitative criteria in the formulation of the hypotheses, the thought experiment has the same validity as the one actually performed. In this way Galileo understood how the Copernican revolution of heliocentrism (the Sun placed at the center of the Solar System and not Earth as medieval claimed instead Scolastica referring to the authority of Aristotle) was correct and how Kepler's laws were correct at an astronomical level.

The scientific method is therefore the way in which science increases the knowledge of Nature and the Universe.

The characteristics of such knowledge are that of being objective, reliable and verifiable.

Inductive method

The scientific method consists of two large macro-sectors.

On the one hand we have the collection of empirical evidence through experiments that must be brought back to a common theoretical logic, on the other we have the hypotheses and theories that must be in agreement with the experimental reality.

This dualism somehow reflects the ancient division of logical reasoning between the inductive method and the deductive method. While Galileo made particular use of the second, Bacon and Newton were frequent users of the first.

Let us briefly see the characteristics of these two different approaches to science and the scientific method and their implications in physical and philosophical terms.

The inductive method was the real driving force of modern physics and only went into crisis many centuries later, when it was clear that the theories formulated were in clear conflict with each other and with the experimental data.

The twentieth century led to a great transformation not only in the theories elaborated, but also in the approach to science, in the philosophical and logical explanation as well as in the method used.

The inductive method starts from empirical observation and ends in the formalization of a theory, carrying out a series of intermediate steps.

Observation identifies the characteristics of the physical phenomenon and measures them with reproducible methods while the subsequent experiment programmed by the observer allows these characteristics to be detected.

After that it is necessary to prepare an analysis of the correlation between the measurements, manipulating the experimental data in order to extract from them the greatest possible content of information.

This correlation is the first step towards the definition of a physical model which must be an abstraction of the real functioning given by the empirical results.

It must be said that the same experiments can lead to different physical models and the goodness of one model, compared to another, is given by the degree of precision with which the experimental data are explained.

The physical model is, in turn, formalized following a mathematical approach for the definition of a mathematical model which contains a series of equations, whose solutions must coincide with the experimental data.

At the end of the cognitive cycle of the inductive method there is the formulation of the theory which, based on the mathematical model, generalizes the physical model and explains the correlation between the measurements and the experimental data.

By applying the inductive method, new knowledge is generated both by abstracting from the particular to the universal as now done, and by subjecting the theory to experimental verification and overcoming it, with the same scheme, if an observation identifies characteristics not in agreement with what is predicted by the theory itself.

This mental scheme was the one applied by Bacon and Newton and which enjoyed considerable success for centuries.

Deductive method

On the other hand Galileo showed himself closer to the deductive method, also called experimental.

The basic idea of the deductive method is that the theory is built at the beginning and not at the end of the cognitive process, as instead happens in the inductive method.

The deductive method starts from the construction of a mathematical theory which determines a physical model from which hypotheses can be formulated; such hypotheses must predict something experimentally measurable.

By carrying out an appropriate experiment, it is observed whether the event foreseen by the theory, and therefore by the hypothesis, occurs or not.

There are two ways of interpreting the verification between experimental observation and theoretical prediction.

For many centuries, it was agreed that the necessary criterion was that of verifiability.

With this criterion it was deduced that, if the coincidence between the prediction of the theory and the experimental reality does not occur, the theory is denied and therefore a new theoretical approach must be formulated. If, on the other hand, the theory agrees with the experimental data, it is correct.

This was the approach given by Galileo himself.

The second way consists instead in the so-called falsificationism, i.e. starting from the assumption that a theory can never be confirmed, but only refuted.

If there is a coincidence between the theoretical forecast and the experimental data, it can simply be concluded that the theory has not been denied and can be accepted on a provisional basis.

This approach derives mainly from Popper's studies during the 20th century.

The deductive method had great impetus after the logical criticisms carried out by Russell in the early twentieth century against the inductive method.

In the meantime, theories and ways of thinking closer to the deductive method developed, such as the studies on relativity carried out by Einstein, and concepts such as probabilism and indeterminism were introduced which sanctioned the definitive decline of induction.

Finally, the enunciation of Godel's incompleteness theorems gave the final blow to that logical scheme, leaving the only way of deduction open.

Popper's studies then made sure that falsificationism was taken as an assumption of today's science.

In particular, Russell posed as a capital point the logical inconsistency of induction which, on the basis of individual cases, abstracted a universal law.

Many contemporary studies tend to confirm this thesis, above all after the evident intrinsic incompleteness of every theory or logical scheme (demonstrated by Godel in the 1920s).

In fact, to make the inductive system valid, it would take an infinite number of empirical cases to confirm it, which would not generate any new knowledge.

Conversely, based only on a limited number of experimental cases, every inductive theory is, in reality, only a conjecture.

As proof of Popper's falsificationism, it must be said that the function of experiments is a refutation, as already observed by Einstein regarding physical theories and the connection with the deductive and experimental method.

Every physical theory can be called scientific if and only if it is expressed in a form that can be criticized and falsified in objective terms. From this point of view, Popper criticized many pseudo-scientific theories such as historicism, psychology, materialism and metaphysics, dismantling, among other things, studies by eminent philosophers such as Marx, Freud, Hegel and Kant.

Applications in physics

Going back to the origins of the scientific method and to Galileo, the first applications of this criterion were stated in 1638, in the scientific treatise *"Mathematical speeches and demonstrations around two new sciences pertaining to mechanics and local motions "*.

This treatise was the dawn of modern physics and that date can be taken as a dividing line between a pre-scientific era and a scientific era.

In that treatise, Galileo generalized the experiments and theories studied in previous years regarding motion on an inclined plane and falling bodies, arriving to correctly describe the laws of statics, leverage and dynamics, especially of motion naturally accelerated, of the uniformly accelerated one and of the oscillatory motion of the pendulum.

Furthermore, Galileo conceived the existence of the vacuum as a state in which there was no resistance of materials and in which motion was possible, arriving correctly to conclude that bodies, having different masses and shapes, fall with equal speed in the vacuum, as opposed to to all the theories of the time.

Always with this approach, Galileo overturned Aristotle's point of view on the principle of inertia through an ideal experiment, i.e. imagining the limiting case of a body moving on a horizontal plane without friction.

In this case, for Galileo, the body remains in its state of motion without any term of space and time, simply for a principle of conservation of energy.

All this knowledge formed the necessary background for the formulation of the laws of Newtonian mechanics in the second half of the seventeenth century, even if there was a need for a new mathematical formulation, that of mathematical analysis, not yet ready in Galileo's time.

Three other aspects of Galileo's scientific method were important for the continuation of modern physics.

The first aspect concerns the astronomical discoveries deriving from the acceptance of the theories of Copernicus and Kepler. Galileo was the first to build a telescope and to scientifically probe celestial objects, such as planets and satellites.

The second aspect is the concept of infinity and its measurement, which will be very useful in mathematical analysis.

The last question concerns the so-called Galilean principle of relativity.

Galileo was the first to scientifically ask himself the question of the validity of physical laws, especially of mechanics, and of the role of different observers in different reference systems.

Galileo started from the hypothesis that the laws of mechanics are always the same for inertial reference systems, ie reference systems that satisfy the principle of inertia. Simply put, such frames of reference are not accelerated.

These reference systems can be expressed through the formalism of the Cartesian axes in three dimensions (with Cartesian coordinates) and by adopting the rules of Euclidean geometry.

The observer present in the reference system is integral with the reference system, therefore it does not have its own motion, but only that of the system.

The first point that Galileo highlighted is that of the simultaneity of the experiment.

Two observers placed in different inertial frames of reference must perform the same experiment at the same instant in order to have an identical result. Therefore they will have to exchange information to synchronize this experiment. Galileo tried to measure the speed of light and deduced that it was so high compared to daily practice, as to make the time necessary for the exchange of information irrelevant.

The first conclusion of Galilean relativity was that time remained the same in the passage from one inertial system to another.

Since the two reference systems have different speeds, Galileo posed the problem of how to carry out a transformation of the speeds, passing from one system to another.

By applying Euclidean geometry together with Cartesian coordinates, he vectorically composed the velocities according to the well-known law of the parallelogram. This law, already known by Leonardo, now found an explanation in the Galilean theory of relativity.

Ultimately, given two inertial systems, the passage of space-time coordinates from one system to another according to Galilean relativity is given by:

$$t' = t$$

$$x' = x - vt$$

Where v is the relative speed between the two systems, composed according to the parallelogram rule.

With these scientific assumptions and with the method developed by Galileo, there were the real foundations for starting the path of modern physics, starting right from the mechanical concepts.

2

MEASURING SYSTEMS

International system: fundamental units

The International System of Measurement (known as SI or MKS System) is a measurement system that is based on the metric system and introduces seven fundamental units for physics.

1) For lengths, the metre, symbol m, is defined as the distance traveled by light in a vacuum in the time of 1/299'792'458 seconds.
2) For the masses, the kilogram, symbol Kg, is defined as the mass of the internationally recognized prototype.
3) For the time, the second, symbol s, is defined as the duration of 9'192'631'770 periods of the radiation corresponding to the transition between two hyperfine levels (from F=4 to F=3 for MF=0) of the state fundamental element of the cesium-133 atom.
4) For the temperature, the Kelvin, symbol K, is defined as 1/273.16 of the thermodynamic temperature of the triple point of water.
5) For the intensity of electric current, the Ampere, symbol A, is defined as the electric current flowing between two linear and parallel conductors, placed in a vacuum at a distance of one meter and producing a force equal to 0.0000002 newton per meter in length.
6) For the amount of substance, the mole, symbol mol, is defined as the amount of substance of a system that contains a number of entities equal to the number of atoms present in 12 grammicarbon-12.
7) For luminous intensity, the candela, symbol cd, is defined as the intensity of a source that emits monochromatic radiation at a frequency of 540 THz with an intensity equal to 1/683 watt per steradian.

International system: derived units

All the others can be derived from the seven fundamental units.
We mention only the main ones:

Frequency	hertz	Hz	$\left[s^{-1}\right]$
Power	newtons	No	$\left[kg \cdot m \cdot s^{-2}\right]$
Pressure	pascal	Pa	$\left[N \cdot m^{-2}\right]$
Power	joule	j	$\left[N \cdot m\right]$
Power	watt	w	$\left[J \cdot s^{-1}\right]$
Electric charge	coulomb	c	$\left[A \cdot s\right]$
Electric potential	volt	v	$\left[W \cdot A^{-1}\right]$
Electric capacity	farad	f	$\left[C \cdot V^{-1}\right]$
Electrical resistance	ohm	Ω	$\left[V \cdot A^{-1}\right]$
Electric conductance	siemens	St	$\left[A \cdot V^{-1}\right]$
Magnetic flux	weber	Wb	$\left[V \cdot s\right]$
Magnetic flux density	tesla	T	$\left[Wb \cdot m^{-2}\right]$
Inductance	henry	h	$\left[Wb \cdot A^{-1}\right]$
luminous flux	lumens	lm	$\left[cd \cdot sr\right]$
Lighting	lux	lx	$\left[lm \cdot m^{-2}\right]$
Radioactive activity	becquerel	Bq	$\left[s^{-1}\right]$
Radioactive dose absorbed	gray	Gy	$\left[J \cdot kg^{-1}\right]$
Equivalent radioactive dose	sievert	St	$\left[J \cdot kg^{-1}\right]$

Recall that in physics, plane angles are always measured in radians and solid angles in steradians.
Also, exponential notations and the use of commas for significant figures are valid.

Prefixes

Being a metric system, the following prefixes apply:

you decide	d	10^{-1}
centi	c	10^{-2}
milli	m	10^{-3}
micro	μ	10^{-6}
dwarf	no	10^{-9}
pic	p	10^{-12}
femto	f	10^{-15}
deed	to	10^{-18}
zepto	z	10^{-21}
yocto	y	10^{-24}
deca	from	10^{1}
hecto	h	10^{2}
kilo	k	10^{3}
mega	m	10^{6}
gig	g	10^{9}
tera	T	10^{12}
peta	P	10^{15}
exa	AND	10^{18}
zetta	Z	10^{21}
yotta	Y	10^{24}

Recall that in computer science there are also base 2 prefixes.

Other commonly used units

Despite the attempt at standardization implemented by the International System, there are some units that lend themselves more to being used, both for common use and for scientific convenience.
We can make the following equivalences:

TIME
Since time is measured in seconds, 1 minute corresponds to 60 seconds, 1 hour to 60 minutes and 1 day to 24 hours.
SPACE
Since lengths are measured in metres, an angstrom equals one-tenth of a nanometer and a nautical mile equals 1' 852 metri.
Since areas are measured in square meters, a hectare equals 10' 000 metri quadratiand a barn equals 100 square femtometres.
Given that volumes are measured in cubic metres, a liter is equivalent to a cubic decimetre.
MASS
Since masses are measured in kilograms, a quintal equals 100 Kgand a ton equals 1' 000 Kg.
SPEED'
Since speeds are measured in meters per second, one knot equals the speed of one nautical mile per hour.
PRESSURES
Since pressures are measured in Pascals, one bar equals 100,000 Pa, one millibar equals 100 Pa, and one millimeter of mercury (mmHg) equals 133.322 Pa.
POWER
Since energies are measured in Joules, one calorie equals 4.1868 J and one kilocalorie equals 4168.8 J.

The following units of measurement are also considered valid:
electronvolt (symbol eV): energy equal to $1.602 \cdot 10^{-19}\,J$
atomic mass unit (symbol u): mass equal to $1.662 \cdot 10^{-27}\,Kg$
astronomical unit (ua): length equal to $1.495 \cdot 10^{11}\,m$

In astronomy, the light year equal to 63'284 au and the parsec equal to 206'625 au are also considered.

CGS system

The CGS system is a derivation of the international system, in which the basic units of kilogram and meter are replaced by those of gram and centimeter.

Other units of measurement of this system that find applications in physics are:

erg: defined as that energy equal to $10^{-7} J$

dyne: defined as that force equal to $10^{-5} N$

poise: defined as that viscosity equal to 0.1Pa*s

British imperial system

It is a system that is not based on the metric system.
We have:

LENGTH:
One inch equals 2,54 cm.
A thousandth of an inch is the mil.
One fortieth of an inch is the line.
One hand equals 4 pollici.
A span equals 9 pollici.
One foot equals 12 pollici.
One elbow equals 18 pollici.
One yard equals 3 piedi.
One arm equals 2 yards.
One pole equals 5.5 yards.
One chain equals 11 fathoms.
One stage equals 10 chains.
One terrestrial mile equals 8 stadia or 1609,344 metri.

MASS:
One ounce equals 28,35 grammi.
An eighth of an ounce is called a dram.
A pound is 16 once.

For areas and volumes, these quantities are used squared and cubed.

We emphasize that one acre equals 0,4046 ettari.
For volumes for liquids, the following equivalences apply.

One fluid ounce equals 28.4 ml.
A pint equals 20 onceliquids.
One gallon equals eight pints.

TEMPERATURES:
The Fahrenheit scale is used.
To switch from Fahrenheit scale to Celsius scale, the following rule is used:

$$°F = (°C \cdot 1{,}8) + 32$$

US customary system

The only differences with respect to the British imperial system consist in the use of the Rankine scale for temperatures with the following equivalence:

$$°R = °F + 459{,}67$$

And a different classification of volumes for liquids:

One fluid ounce equals 29.57 ml.
A pint equals 16 onceliquids.
One gallon equals eight pints.

3

CLASSICAL MECHANICS: KINEMATICS

Definitions

Kinematics deals with the study of the so-called time law, i.e. the relationship that exists between the distance covered by a point as time varies.

The kinematics does not ask itself the question of what is the origin of this motion, whose task is left to the dynamics.

To establish any time law, some concepts must be defined.

Starting from Newton's hypotheses, space and time are absolute quantities that express physically measurable coordinates through human experience.

Velocity is defined as the ratio of space to elapsed time.

We speak of average speed if referred to an interval of time and instantaneous speed if referred to a single instant.

Acceleration is defined as the ratio of velocity times elapsed time.

We speak of average acceleration if referred to an interval of time and instantaneous acceleration if referred to a single instant.

In formulas we have:

$$v_m = \frac{\Delta s}{\Delta t}$$

$$v_{ist} = \lim_{\Delta t \to 0} \frac{\Delta s}{\Delta t}$$

$$a_m = \frac{\Delta v}{\Delta t}$$

$$a_{ist} = \lim_{\Delta t \to 0} \frac{\Delta v}{\Delta t}$$

If a motion has zero velocity, it can be seen that the point is at rest.

If a motion has constant speed, it is called uniform motion.

In uniform motion, acceleration is zero.

If a motion has constant acceleration, it is said to be uniformly accelerated. In uniformly accelerated motions, speed increases proportionally with time.

Distance, velocity and acceleration are defined as vector quantities and not as scalar quantities.
A vector is a mathematical object characterized by three parameters: the module (ie the numerical value that coincides with the scalar), the directions and the direction.
A vector can be seen as an oriented segment in the Cartesian plane (or space).
Therefore, to understand the kinematics knowledge of analytical geometry of the Cartesian plane is necessary, which we take here for granted.
Motion in which the space vector does not change direction is called rectilinear motion.
Conversely, the motion is called curvilinear.
Compound motion is a motion given by the union of two simple motions in its directions (for example a two-dimensional motion in which there is uniform motion in one dimension and uniformly accelerated motion in the other).

One last note before continuing.
Frictions are not considered in kinematics, i.e. those phenomena which oppose the real motion of the points (such as, for example, the air resistance during the fall of an object).
For a treatment of the frictions it is necessary to resort to the dynamics.

Uniform rectilinear motion

Uniform rectilinear motion is characterized by zero acceleration and constant velocity.
Therefore the timetable will simply be:

$$s(t) = vt$$

If the object does not start from a point considered to be zero (the origin of the Cartesian axes), the previous equation transforms into:

$$s(t) = vt + s_0$$

We note how uniform rectilinear motion is characterized by a straight line in the Cartesian plane, where the x axis coincides with time and the y axis with space.

Uniformly accelerated motion in a straight line

In uniformly accelerated rectilinear motion, the acceleration is constant, therefore the speed follows a direct proportionality with the time:

$$v(t) = at$$

The hourly law of this motion is given by:

$$s(t) = \frac{1}{2} at^2$$

Knowledge of mathematical analysis is required to introduce the factor ½ (refer to the last chapter of this manual).
Let's see how the time law in the Cartesian plane is represented by a parabola with its vertex at the origin of the axes.
This motion physically represents the fall of a grave.
If we consider an object left free to fall from a certain height, the acceleration will be equal to that of gravity and, from the hourly law, we can obtain the time necessary for the object to touch the ground.
Given h the height at which it is placed and g the acceleration due to gravity, we simply have that:

$$t = \sqrt{\frac{2h}{g}}$$

The speed with which it hits the ground is given by:

$$v = \sqrt{2gh}$$

If the body has zero velocity at the instant t=0 the above equations are modified in this way:

$$v(t) = at + v_0$$

$$s(t) = \frac{1}{2}at^2 + v_0 t$$

Finally, if the body does not start from a space that we can denote as origin we have:

$$v(t) = at + v_0$$

$$s(t) = \frac{1}{2}at^2 + v_0 t + s_0$$

Which is the most general form of the uniformly accelerated motion law.
Graphically, the hourly law is any parable.
With this formalism we can calculate the maximum height that a body reaches if thrown upwards at a certain initial speed.
In fact, the body will undergo a uniformly decelerated motion due to the action of gravity.
The body will stop after a time given by:

$$t = \frac{v_0}{g}$$

Reaching a height equal to:

$$h = \frac{1}{2}g\frac{v_0^2}{g^2} + \frac{v_0^2}{g} = \frac{3}{2}\frac{v_0^2}{g}$$

A special case of uniformly accelerated motion is motion along an inclined plane.
Given an inclined plane of length l and height h, we can simply derive the acceleration to which the body is subjected when falling from the inclined plane.
With simple goniometry considerations it can be seen that the acceleration is given by:

$$a = g \cdot \sin \alpha$$

Where alpha is the angle of inclination of the plane.
For those who do not have knowledge of goniometry, it is always possible to derive a formula which links the height h and the length l from the similarities of right triangles (but, in this case, not knowing the goniometry,

it will be possible to solve the problem only for right triangles known from plane geometry, such as those with base angles of 30°, 45° and 60°).
The discourse proceeds in the same way as for the fall of a grave.

Compound rectilinear motions

A compound rectilinear motion is a motion given by the superimposition of two simple motions along the axes.
For the sake of simplicity, let us consider two-dimensional motions.
A rectilinear motion composed of two uniform motions is simply a uniform rectilinear motion whose direction is given by the vector sum of the motions along the two directions.
This motion is solved in a trivial way with the parallelogram rule for vector sums.
A rectilinear motion composed of a uniform motion and a uniformly accelerated motion gives rise to the study of two very interesting practical cases.
The first is the fall of a body after it has undergone a uniform motion, for example we can think of a marble that rolls on a table at a constant speed and then, at the end of the table, is subjected to the fall due to the motion of gravity .
The trajectory it will draw will be a parabola with the vertex at the end of the table and the concavity facing downwards (since gravity is directed downwards).
The more speed you gain while sliding across the table, the farther the ball will fall from the edge of the table.
A variant of this physical problem is given by the motion of a projectile, particularly important in ballistics for the range of cannons.
The motion of a projectile can be summarized as uniform motion along the vertical axis (with known initial velocity) and uniformly decelerated motion along the horizontal axis.
To increase the muzzle velocity, the only technological way adopted was to increase the diameters and lengths of the guns (hence the reason for the calibers and lengths of the gun barrels).
It can easily be seen that the maximum reachable distance, called range, is obtained for an inclination of 45° (in jargon it is called launch angle).
It can be seen from the Pythagorean theorem that, given an initial velocity, its components broken down along the x-axis and the y-axis by an angle of 45° are given by:

$$v_{0x} = v_{0y} = \frac{v_0}{\sqrt{2}}$$

The time required for the projectile to reach maximum height is given by:

$$t_1 = \frac{v_{0y}}{g} = \frac{v_0}{g\sqrt{2}}$$

The time required for the projectile to reach the ground again is given by:

$$t = 2t_1 = \frac{\sqrt{2}v_0}{g}$$

And the range will be given by:

$$s = 2s_1 = \frac{v_0^2}{g}$$

The maximum height reached will be:

$$h = \frac{v_0^2}{2g}$$

In the case of any angles, we have to resort to goniometry.
The maximum range and height are respectively given by:

$$s = \frac{v_0^2 \sin 2\alpha}{g}$$

$$h = \frac{v_0^2 \sin^2 \alpha}{2g}$$

where alpha is the launch angle.
As already mentioned, air friction and other factors (such as wind direction and speed and the earth's rotation) which are absolutely fundamental in matters of ballistics are omitted.

Uniform circular motion

A motion in which the space vector changes direction is not rectilinear.
Among non-rectilinear motions, circular motion has particular interest, i.e.
the motion that describes a circumference.
Let us study the simple case of uniform circular motion, i.e. at constant
speed.
This case exemplifies, for example, the motion of a point on a wheel that
rotates at a constant speed.
In a circular motion, the time law is defined on the angles and not on the
space.
This arises from the geometric nature of the motion.
As time passes, the angle covered in the center of the circumference will
be gradually greater.
The horary law will correlate the angle with a speed called angular:

$$\vartheta(t) = \omega t$$

If the angle does not start from zero we have:

$$\vartheta(t) = \omega t + \vartheta_0$$

Angular velocity is measured in radians per second.
It is this speed that is constant and not the speed of the single point P
placed on the circumference.
The speed of the point P is associated with the name of tangential speed,
defined as follows:

$$v = R\omega$$

Where R is the radius of the circle.
As we can see, points belonging to concentric circles that rotate at the
same angular speed have different tangential speeds.
The tangential speed is a vector having a tangent direction to the
circumference at point P and a direction equal to the direction of rotation
of the circular motion.

Uniform circular motion has two characteristics which distinguish it in a
peculiar way from rectilinear motions.
The first is that, although it is a uniform motion, the acceleration is zero.
There is an acceleration whose value is given by:

$$a = \frac{v^2}{R} = R\omega^2$$

The acceleration is a vector having for module the scalar just exposed, for direction the one joining the point P with the center of the circumference and towards the center.

This is why we speak of centripetal acceleration.

The second peculiarity derives from the fact that this motion is periodic.

In fact, after a rounded corner, point P returns on itself.

The period will therefore be given by:

$$T = \frac{2\pi}{\omega}$$

Other circular motions

We only mention the existence of uniformly accelerated circular motion, whose time law is given by:

$$\vartheta(t) = \vartheta_0 + \omega_0 t + \frac{1}{2}\alpha t^2$$

where alpha is the angular acceleration and is the value that remains constant over time.

Furthermore, even for circular motions we can consider compound motions.

For example, a uniform circular motion in the Cartesian plane combined with a uniform rectilinear motion in the third spatial dimension causes a helical-cylindrical motion.

Harmonic motion

Harmonic motion plays a fundamental role in physics and that is why we dedicate this chapter to it.

We can derive the harmonic motion starting from the uniform circular motion

Said P a point that moves on a circumference of uniform circular motion, what kind of motion will the points A and B have given by the projections of P along the Cartesian axes?

The motion of A and B are harmonic motions whose general time law is as follows:

$$x(t) = A \cos(\omega t + \phi)$$

where A is a constant defined as amplitude, the angle that adds up to the time term is called phase, while omega is called pulsation.

In the case of uniform circular motion, it can be seen that the projections of P on the axes have trends equal to the cosine function for the projection on the abscissa axis and sine for the function on the ordinate axis.

Therefore a uniform circular motion can be seen as the superposition of two harmonic motions along the coordinate axes:

$$x(t) = A \cos(\omega t)$$
$$y(t) = A \sin(\omega t)$$

The amplitude A coincides with the radius of the circumference of the circular motion and the omega pulsation coincides with the angular velocity of the circular motion.

Therefore also the period of a harmonic motion is the same as that of the uniform circular motion.

For those familiar with goniometry it is easy to understand this parallelism, as it is a rewriting of the fundamental relationship of goniometry:

$$\sin^2(\omega t) + \cos^2(\omega t) = 1$$
$$A = R = 1$$

For those unfamiliar with goniometry, let's say that along the x axis, the projection point of P moves from the value assumed for an angle equal to zero to the center of the circumference (for an angle = 90°) and then moves towards the negative abscissas up to the side opposite (for angle = 180°) and from here retrace your steps.

On the y axis instead there will be a behavior that takes the point from the center (angle = 0°) to the maximum positive extreme (angle = 90°) and then brings it back to the center (angle = 180°), to the maximum negative extreme (angle = 270°) and back to center for angle = 360°.

The speed and accelerations of a harmonic motion are also expressed by trigonometric functions, in particular we have:

$$v(t) = \omega A \cos\left(\omega t + \phi + \frac{\pi}{2}\right)$$

$$a(t) = \omega^2 A \cos(\omega t + \phi + \pi)$$

Returning to the case of circular motion, we have that the speed is maximum when the projection points of the point P pass through the center and is zero at the extremes (where the motion is reversed).
Acceleration, on the other hand, is maximum at the extremes and zero at the centre.

The interest in harmonic motion derives from the hundreds of physical applications it has.
Harmonic motion can be used to describe the following situations:
- oscillatory motion of a spring
- motion of a pendulum
- oscillatory phenomena in electric circuits

We must always emphasize that the discourse always holds in the absence of friction.
Harmonic motion describes these situations and determines the periods of oscillation, such as

Oscillating spring without damping

$$T = 2\pi\sqrt{\frac{m}{C}}$$

$$F = C\Delta l$$

Physical pendulum

$$T = 2\pi\sqrt{\frac{I}{\tau}}$$

Torsion pendulum

$$T = 2\pi\sqrt{\frac{I}{k}}$$

$$k = \frac{2lm}{\pi r^4 \Delta\varphi}$$

Math pendulum

$$T = 2\pi \sqrt{\frac{l}{g}}$$

I indicates the moment of inertia, k the torsion constant, τ the moment of force, g the acceleration due to gravity, l the length of the pendulum.
Finally, we underline how A is the amplitude of the oscillations (hence the name of the constant).

Derivation from Newtonian mechanics

In 1686 Isaac Newton published the *"Mathematical principles of natural philosophy"* , a work that can be considered the real beginning of modern physics.
The leap in quality compared to Galileo's times was given by the introduction of mathematical analysis as a tool for expressing physical equations.
The formulation of the first foundations of mathematical analysis is a merit Newton must share with a philosopher like Leibnitz, even if at the time there was heated controversy about who was the first to identify this evolution of mathematics.
In that paper Newton laid the foundations of classical mechanics, the first physical discipline to be probed in depth with the scientific method.
First, he defined the first laws of kinematics starting from considerations of mathematical analysis.
Considering a Cartesian coordinate system, Newton defined velocity and acceleration as first and second derivatives of space with respect to time.

$$\bar{v} = \frac{d\bar{s}}{dt} = s'(t) = \dot{s}$$

$$\bar{a} = \frac{d\bar{v}}{dt} = \frac{d^2\bar{s}}{dt^2} = v'(t) = \dot{v} = s''(t) = \ddot{s}$$

In these equations we have summarized the physical, mathematical and mechanical notations of the derivatives. Both space and velocity and acceleration are vector quantities.

For a spherical coordinate system instead we can define the radius of curvature as:

$$\rho = \frac{1}{|k|} = \left(\frac{d^2\vec{r}}{ds^2} \right)^{-1}$$

In this case the speed and acceleration become the angular ones, given by:

$$\omega = \dot{\vartheta}$$

$$\alpha = \ddot{\vartheta}$$

The transition from spherical to Cartesian coordinates is implemented by applying the well-known Leibnitz rule for products of derivatives.

By doing so, we are able to calculate the equations of motion, at least in simple cases.

For uniform motion in a straight line, the acceleration is identically equal to zero, the velocity is constant, and the equation of motion is simply given by

$$s(t) = vt + s_0$$

For uniformly accelerated motion in a straight line, the equations of motion are:

$$v(t) = v_0 + at$$

$$s(t) = s_0 + v_0 t + \frac{1}{2} at^2$$

This motion also describes the fall of a body, setting the acceleration equal to that of gravity.

The acceleration of gravity assumes, on the Earth, an almost constant value, considering the variations due to the non-perfect sphericity of our planet and to the height of the place with respect to the sea level (completely negligible with respect to the radius of the Earth) irrelevant.

This formulation foresees and confirms what Galileo had deduced about the fall of bodies in a vacuum.

With simple geometric considerations, it is also possible to include in this equation everything connected to the motion of falling along an inclined plane.

For circular motions, analogous considerations apply simply referring to speed and angular acceleration and not as regards rectilinear motions (called tangential motions in this case).
These equations are valid in three-dimensional space and can be used to describe more complex cases, such as those of combined motions.
A typical example is given by ballistics studies: a projectile exploded from a cannon will have uniform rectilinear motion along the horizontal axis and uniformly decelerated motion along the vertical one, thus generating a compound motion that exactly describes the real parabola of this object .

High school exercises

Exercise 1

A vehicle travels for a certain time T at a speed of 40 km/h, covering a given distance d.
After that, it travels the same distance but at 80 km/h.
Find the average speed.

The distance traveled is equal to 2d.
The total time is given by:

$$\Delta t = d/v_1 + d/v_2.$$

So the average speed is:

$$v_m = \frac{\Delta s}{\Delta t} = \frac{2d}{\frac{d}{v_1} + \frac{d}{v_2}} = 2\frac{v_1 v_2}{v_1 + v_2} \simeq 53.3 \text{ km/h} \equiv 14.8 \text{ m/s}.$$

Exercise 2

An object is thrown upwards from the ground with an initial speed of 12 m/s.

31

How long does it take to reach the highest point?
What is the height of the highest point?
How long after the launch does it fall to the ground?
How fast does it hit the ground?

We neglect friction and apply the laws of kinematics.
We have:

$$v(t) = v_0 - g\,t$$

The time to reach the highest point is found when v(t)=0 i.e.:

$$t_{max} = \frac{v_0}{g} = 1.24 \text{ s};$$

The distance covered in this time coincides with the maximum height reached:

$$d = v_0 t_{max} - \tfrac{1}{2} g\, t_{max}^2 = 7.3 \text{ m};$$

The time after which the object returns to the ground is simply twice the time to reach the highest point since, in the absence of friction, the forward path is identical to the return path, except considering the appropriate signs for the vectors space and speed:

$$t_{terra} = 2t_{max} = 2.48 \text{ s};$$

From which:

$$v_{terra} = -v_0 = -12 \text{ m/s}.$$

Exercise 3

Two objects are 20 km apart in a one-dimensional line.
They leave at the same instant going towards the other object, the first with a constant speed of 50 Km/h, the second of 100 Km/h.
How soon do they meet?

The timetables are:

$$X_1 = v_1 t \qquad\qquad X_2 = v_2 t$$

Furthermore, it must be valid:

$$X_1 + X_2 = v_1 t + v_2 t = 20 \text{ km}$$

Transferring to the international system, we have:

$$v_1 = 50 \text{ km/h} = 13.89 \text{ m/s}$$
$$v_2 = 100 \text{ km/h} = 27.778 \text{ m/s}$$
$$20\,000 \text{ m} = (13.89 \text{m/s}) \, t + (27.778 \text{ m/s}) \, t$$
$$t = 20\,000 / 41.67 \text{ s} = 480 \text{ s}$$

Exercise 4

A particle has the following time law:

$$9.75 + 1.50t^3$$

Considering the interval between 2 and 3 seconds, calculate the average speed, instantaneous speeds at 2 and 3 seconds, the same amounts when halfway.

After 2 seconds the particle will have traveled:

$$x(2) = 9.75 + 1.50 \cdot 2^3 = 21.75 m$$

After 3 seconds:

$$x(3) = 9.75 + 1.50 \cdot 3^3 = 50.25 m.$$

The average speed is equal to the difference between 50.25 and 21.75 divided by the difference between 3 and 2 or 28.5 m/s.
The instantaneous speeds will be:

$$v_{ist}(2) = 4.50 \cdot 2^2 = 18 \tfrac{m}{s}$$

33

$$v_{ist}\,(3) = 4.50 \cdot 3^2 = 40.5\,\tfrac{m}{s}$$

Half way corresponds to x=36 m.
We have:

$$t = \sqrt[3]{\frac{36 - 9.75}{1.5}} = 2.6\,s$$

And then:

$$v_{ist}\,(2.6) = 4.50 \cdot 2.6^2 = 30.3 \cdot \tfrac{m}{s}$$

Exercise 5

An object starts from rest and moves with constant acceleration.
At one instant it is traveling at 30m/s, after 160m it is traveling at 50m/s.
Calculate the acceleration, the time taken to travel 160 m, the time taken to
reach 30 m/s and the distance traveled from the start until the speed of 30
m/s is reached.

The motion is uniformly accelerated so:

$$v_f^2 = v_i^2 + 2a\,\triangle x$$

Therefore, the acceleration is:

$$a = \frac{v_f^2 - v_i^2}{2\,\triangle x} = \frac{\left(50^2 - 30^2\right)\frac{m^2}{s^2}}{2 \cdot 160\,m} = 5\,\frac{m}{s^2}$$

From the definition of acceleration:

$$a = \frac{v_f - v_i}{t_f - t_i}$$

We get the required time:

$$\Delta t = \frac{v_f - v_i}{a} = \frac{(50 - 30) \frac{m}{s}}{5 \frac{m}{s^2}} = 4\,s$$

The time taken to reach 30 m/s from standstill is:

$$\Delta t = \frac{(30 - 0) \frac{m}{s}}{5 \frac{m}{s^2}} = 6\,s$$

And the distance traveled in this time frame:

$$s = \frac{1}{2} a t^2 = \frac{1}{2} \cdot 5a = \frac{m}{s^2} \cdot 6^2 s^2 = 90\,m$$

Exercise 6

An object is thrown downwards with a speed of 12 m/s from a height of 30 meters.
How long does it take to get to the ground? How fast does it hit the ground?

It is a uniformly accelerated motion whose time law, in the absence of friction, is:

$$s = v_0 t + \tfrac{1}{2} g t^2$$

Substituting the values, we have:

$$30m = 12.0 \frac{m}{s} \cdot t + 4.9 \frac{m}{s^2} \cdot t^2$$

From which:

$$t_{1,2} = \frac{-12 \pm \sqrt{144 + 120 \cdot 4.9}}{9.8} = 1.53\,s$$

Where the negative solution was discarded as having no physical meaning.
The final speed will be given by:

$$v_f^2 = v_0^2 + 2gh;$$

And then:

$$v_f^2 = (12.0)^2 \frac{m^2}{s^2} + 2 \cdot 9.8 \frac{m}{s^2} \cdot 30m = 27 \frac{m}{s}$$

Exercise 7

An object travels at a constant speed of 60 km/h for 40 minutes due east, then for 20 minutes in a direction forming an angle of 50° due east to north, and finally due west for 50 minutes.
What is its average vector speed over the entire journey?

The problem can be schematized as:

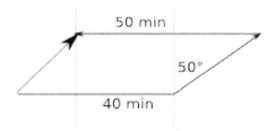

The distance covered in the various sections is:

$$s_1 = 60.0 \, \frac{km}{h} \cdot \frac{2}{3}h = 40 \, km$$

$$s_2 = 60.0 \, \frac{km}{h} \cdot \frac{1}{3}h = 20 \, km$$

$$s_3 = 60.0 \, \frac{km}{h} \cdot \frac{5}{6}h = 50 \, km$$

Considering the components of the displacement of the second section along the north-south (x-axis) and east-west (y-axis) directions, we have:

36

$$s_{2x} = 20 \cdot \sin 50 = 15.3 \, km$$
$$s_{2y} = 20 \cdot \cos 50 = 12.9 \, km$$

While the first segment and the third segment are exclusively lying in the x direction (the first segment is positive, the third negative).
The vector sum is given by:

$$s_x = 40 - (50 - 15.3) = 5.3 \, km$$
$$s_y = s_{2y} = 12.9$$

The magnitude of the displacement vector and the angle are:

$$s = \sqrt{(5.3)^2 + (12.9)^2} = 13.9 \, km$$

$$\alpha = \arctan \frac{12.9}{15.3} = 40.1°$$

The required average speed is:

$$v_{media} = \frac{13.9 \, km}{\frac{11}{6} h} = 7.7 \, \frac{km}{h}$$

Exercise 8

An object departs from the origin at time t=0 with an initial velocity given by:

$$\vec{v} = 8.0 \vec{j} \, \frac{m}{s}$$

The object moves in the xy plane with a constant acceleration equal to:

$$\vec{a} = 4.0 \vec{i} + 2.0 \vec{j} \, \frac{m}{s^2}.$$

Find the value of the y coordinate when x=29.
Find the value of the scalar velocity at that instant.

The vector hourly law of motion is:

$$\vec{s} = \vec{v}_0 t + \frac{1}{2}\vec{a}t^2$$

Which is divided into:

$$s_x = v_{0x}t + \frac{1}{2}a_x t^2$$

$$s_y = v_{0y}t + \frac{1}{2}a_y t^2$$

If x=29, knowing that the velocity component in y is zero, we have:

$$29 = 2.0 \cdot t^2$$

From which:

$$t = \sqrt{\frac{29}{2.0}} = 3.8\,s$$

So the spatial coordinate in y will be:

$$s_y = 8.0 \cdot 3.8 + 1.0 \cdot 3.8^2 = 45\,m$$

Being:

$$\vec{v} = \vec{v}_0 + \vec{a}t$$

I our case:

$$v_x = 4.0 \cdot 3.8 = 15.2\,\tfrac{m}{s}$$

$$v_y = 8.0 + 2.0 \cdot 3.8 = 15.6\,\tfrac{m}{s}$$

And so the scalar velocity is:

$$v = \sqrt{15.2^2 + 15.6^2} = 22\,\frac{m}{s}$$

Exercise 9

A rifle is aimed horizontally at a target 30 meters away. Bullet strikes target 1.9cm below center.
Determine the time of flight of the projectile and its initial speed.

The motion is of the descending parabolic type as in the figure:

The motion is composed of uniform rectilinear motion along x and uniformly accelerated rectilinear motion along y.
The velocity component along the y axis is zero.
The only acceleration is given by that of gravity.
We have:

$$y - y_0 = 1.9 \cdot 10^{-2}\,m = 4.9\,\frac{m}{s^2} \cdot t^2\,s^2$$

And so the flight time is:

$$t = \sqrt{\frac{1.9 \cdot 10^{-2}\,m}{4.9\,\frac{m}{s^2}}} = 0.062\,s$$

The initial speed depends only on the component in x which is given by a uniform motion:

$$v = \frac{30\,m}{0.062\,s} = 482\,\frac{m}{s}$$

Exercise 10

An object falls horizontally from a height of 1.2 meters.
It is found that the object falls to the ground 1.5 meters away from the edge of the height.
Calculate the time of flight of the object and its initial speed.

The motion is of the descending parabolic type.
The motion is composed of uniform rectilinear motion along x and uniformly accelerated rectilinear motion along y.
The velocity component along the y axis is zero.
The only acceleration is given by that of gravity.
We have:

$$t = \sqrt{\frac{2h}{g}} = \sqrt{\frac{2.40\,m}{9.81\,\frac{m}{s^2}}} = 0.50\,s$$

The initial speed depends only on the component in x which is given by a uniform motion:

$$v = \frac{\Delta s}{\Delta t} = \frac{1.50\,m}{0.5\,s} = 3.0\,\frac{m}{s}$$

Exercise 11

What is the maximum height that an object can reach if it is thrown from a system with a maximum range of 60 meters?

The motion is parabolic and the maximum height is the vertex of the parabola.

40

It will be:

$$H = \frac{v_0^2 \sin^2 \vartheta_0}{2g}$$

While the range is:

$$R = \frac{v_0^2 \sin 2\vartheta_0}{g}$$

The range is maximum for an angle of 45° and becomes:

$$R = \frac{v_0^2}{g}$$

Substituting into the height formula, we get:

$$H = \frac{R \sin^2 45°}{2} = \frac{60.0 \, m \left(\frac{\sqrt{2}}{2}\right)^2}{2} = 15.0 \, m$$

Exercise 12

Determine the magnitude, direction and direction of acceleration of an object moving at 10 m/s on a curve of radius 25 meters.

It is a uniform circular motion.
The acceleration is centripetal with a direction equal to the one joining the center of the circumference with the point and towards the center of the circumference.
The modulus of this acceleration is:

$$a_c = \frac{v^2}{r} = \frac{\left(10 \, \frac{m}{s}\right)^2}{25 \, m} = 4 \, \frac{m}{s^2}$$

Exercise 13

A rotating object makes 1,200 revolutions per minute. Consider a point on the outer edge of this object (which has a radius of 0.15m).
Find the distance this point travels in each revolution, its speed and its acceleration.

It is a uniform circular motion.
The frequency of the motion is:

$$f = \frac{1200 \frac{giri}{min}}{60 \frac{s}{min}} = 20\, Hz$$

The distance that the point travels in each revolution is the measure of the circumference:

$$l = 2\pi r = 2\pi \cdot 0.15\, m = 0.94\, m$$

The speed is given by:

$$v = 2\pi r f = 0.94\, m \cdot 20\, Hz = 18.8\, \frac{m}{s}$$

And the acceleration from:

$$a = \frac{v^2}{r} = \frac{18.8^2 \frac{m^2}{s^2}}{0.15\, m} = 2369\, \frac{m}{s^2}$$

Exercise 14

A train travels at 216 km/h.
Knowing that the maximum acceleration tolerated by the passengers is 5% of the earth's gravity, determine the minimum allowable radius for the curves of the tracks.
If a curve has a radius of 1 km, how far will the train have to go in these conditions?

An acceleration of 5% of the earth's gravity corresponds to:

$$0.050 \cdot 9.8 \, \tfrac{m}{s^2} = 0.49 \, \tfrac{m}{s^2}.$$

The speed of the train in the international system is:

$$v = \tfrac{216}{3.6} \, \tfrac{m}{s} = 60 \, \tfrac{m}{s}.$$

The radius of curvature is found:

$$r = \frac{v^2}{a} = \frac{60^2 \, \frac{m^2}{s^2}}{0.49 \, \frac{m}{s^2}} = 7347 \, m$$

Conversely, if the radius is 1000 meters, the maximum speed will be:

$$v = \sqrt{ar} = \sqrt{0.49 \, \frac{m}{s^2} \cdot 1000 \, m} = 22.1 \, \frac{m}{s} = 80 \, \frac{km}{h}$$

University-level exercises

Exercise 1

The acceleration of a point P moving along the x axis is:

$$a(x) = (3x - 1) \, \mathrm{m\,s^{-2}}.$$

The point has a given initial velocity.
Study the function v(x).

Recalling that:

$$\frac{1}{2}v^2 - \frac{1}{2}v_0^2 = \int_{x_0}^{x} a(x') \, \mathrm{d}x' \, .$$

We have:

$$v^2 = v_0^2 + 2\int_0^x (3x' - 1)\, dx' = v_0^2 + 3x^2 - 2x .$$

$$v(x) = \pm \sqrt{3x^2 - 2x + v_0^2} .$$

Based on the value of the initial speed, there are three cases:

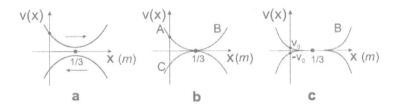

Case a occurs if:

$$v_0^2 > 1/3 \, \mathrm{m^2\, s^{-2}}$$

Case b if:

$$v_0^2 = 1/3 \, \mathrm{m^2\, s^{-2}}$$

Case c if:

$$v_0^2 < 1/3 \, \mathrm{m^2\, s^{-2}} \; |$$

Exercise 2

A point moves as follows:

$$x = t^2 ,$$
$$y = (t - 1)^2 ,$$

Determine the trajectory and study the trend of speed, acceleration and radius of curvature.
We eliminate the parameter t:

$$\begin{cases} t^2 = x \\ y = t^2 - 2t + 1 \end{cases} \qquad \begin{cases} t = \pm\sqrt{x} \\ y = x \pm 2\sqrt{x} + 1 \end{cases}$$

From which:

$$y^2 + x^2 - 2xy - 2y - 2x + 1 = 0 .$$

The trajectory is a conic.
By making this change of coordinates:

$$\begin{aligned} x &= x'\cos\theta + y'\sin\theta = \sqrt{2}x'/2 + \sqrt{2}y'/2 \\ y &= -x'\sin\theta + y'\cos\theta = -\sqrt{2}x'/2 + \sqrt{2}y'/2 \end{aligned}$$

We have:

$$y' = \frac{\sqrt{2}}{2} x'^2 + \frac{\sqrt{2}}{2} .$$

Therefore it is a parabola with the axis coinciding with the bisector of the I and III quadrants:

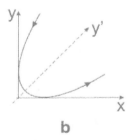

b

For the speed we have:

$$v_x = 2t , \qquad\qquad v_y = 2t - 2 .$$

So:

$$v = \sqrt{v_x^2 + v_y^2} = \sqrt{8t^2 - 8t + 4} \, .$$

For acceleration:

$$a_x = 2 \, , \qquad a_y = 2 \, .$$

From which:

$$a = a_T \, \hat{u}_T + a_N \, \hat{u}_N \, ,$$

$$a_T = \frac{dv}{dt} = \frac{4t - 2}{\sqrt{2t^2 - 2t + 1}} \, ;$$

$$a_N = \sqrt{a_x^2 + a_y^2 - a_T^2} = \frac{2}{\sqrt{2t^2 - 2t + 1}} \, .$$

Since:

$$a_N = v^2 / \varrho.$$

The radius of curvature is:

$$\varrho = \frac{v^2}{a_N} = 2 \left(2t^2 - 2t + 1\right)^{3/2} \, .$$

Exercise 3

Given:

Whose time law is:

46

$$r(t) \ = \ \hat{\imath} \, A \cos(\omega t) \ + \ \hat{\jmath} \, B \sin(\omega t) \, ,$$

Determine the equations of the trajectory of P.
Study the speed and acceleration of P.

The coordinates of point P are:

$$x \ = \ A \ \cos(\omega t) \, ,$$
$$y \ = \ B \ \sin(\omega t) \, .$$

Substituting:

$$\begin{cases} x^2 \ = \ A^2 \ \cos^2(\omega t) \\ y^2 \ = \ B^2 \ \sin^2(\omega t) \end{cases} \qquad \begin{cases} x^2/A^2 \ = \ \cos^2(\omega t) \\ y^2/B^2 \ = \ \sin^2(\omega t) \end{cases}$$

And finally:

$$x^2/A^2 \ + \ y^2/B^2 \ = \ 1 \, .$$

The trajectory is therefore an ellipse, as in the figure:

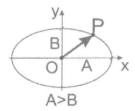

A>B

Speed is:

$$v \ = \ \frac{\mathrm{d}r}{\mathrm{d}t} \ = \ - \, \hat{\imath} \, A\omega \sin(\omega t) \ + \ \hat{\jmath} \, B\omega \cos(\omega t) \, .$$

Or:

47

$$v = -\,\hat{\imath}\,\frac{A}{B}\,\omega y + \hat{\jmath}\,\frac{B}{A}\,\omega x \, .$$

Velocity is always tangent to the trajectory.
The form is:

$$v = \sqrt{v_x^2 + v_y^2} = \omega\,\sqrt{\frac{A^2}{B^2}\,y^2 + \frac{B^2}{A^2}\,x^2} \, .$$

The acceleration is:

$$a = \frac{\mathrm{d}v}{\mathrm{d}t} = -\,\hat{\imath}\,A\omega^2\cos(\omega t) - \hat{\jmath}\,B\omega^2\sin(\omega t) \, .$$

Or:

$$a = -\,\hat{\imath}\,\omega^2 x - \hat{\jmath}\,\omega^2 y = -\omega^2\left(\hat{\imath}\,x + \hat{\jmath}\,y\right) = -\omega^2\,r \, .$$

The acceleration is centripetal of magnitude equal to:

$$a = \omega^2 r.$$

We have:

$$a = a_T\,\hat{u}_T + a_N\,\hat{u}_N \, .$$

Where is it:

$$a_T = \frac{2\omega^2\,xy\,(A^2 - B^2)}{AB\,\sqrt{B^2 x^2/A^2 + A^2 y^2/B^2}} \, .$$

$$a_N = v^2/\varrho \, ,$$

Exercise 4

A point P moves with the clockwise law:

$$x(t) \;=\; A \; \sin(\omega t) \,, \qquad\qquad y(t) \;=\; A \; \cos(\omega t) \,,$$

Find trajectory, speed and acceleration.

We have:

$$x^2 \;=\; A^2 \; \sin^2(\omega t)$$
$$y^2 \;=\; A^2 \; \cos^2(\omega t)$$

Therefore:

$$x^2 \;+\; y^2 \;=\; A^2 \,.$$

The path is a circle of radius A.
For speed:

$$\begin{cases} v_x \;=\; dx/dt \;=\; A\omega \cos(\omega t) \,, \\ v_y \;=\; dy/dt \;=\; -\, A\omega \sin(\omega t) \,. \end{cases}$$

Therefore:

$$v \;=\; \sqrt{v_x^2 + v_y^2} \;=\; A\omega \;=\; \text{costante} \,.$$

So the motion is uniform circular.
For acceleration:

$$\begin{cases} a_x \;=\; dv_x/dt \;=\; -A\omega^2 \sin(\omega t) \,, \\ a_y \;=\; dv_y/dt \;=\; -A\omega^2 \cos(\omega t) \,. \end{cases}$$

Therefore:

$$a \;=\; \sqrt{a_x^2 + a_y^2} \;=\; \omega^2 A \,.$$

Acceleration is centripetal.

Exercise 5

A wheel of radius R rolls without slipping on a horizontal plane.
Given the linear velocity of the center C and the angular velocity with respect to the center, determine the equation of the trajectory of a point P placed at a distance r from the center. (r<R).
Study the speed and acceleration of P.

A point Q on the edge of the wheel has:

$$v_Q = v'_Q + v_0 = 0 .$$

But:

$$v_0 = \omega R .$$

So the time law for the center is:

$$x_c = v_0 t = \omega R t ,$$
$$y_c = R .$$

And those of point P are:

$$x_p = x_c - r \sin \theta = R\omega t - r \sin(\omega t) ,$$
$$y_p = y_c - r \cos \theta = R - r \cos(\omega t) .$$

The trajectory curve is a cycloid.
For the speed we have:

$$v_x = \omega R - \omega r \cos(\omega t) ,$$
$$v_y = \omega r \sin(\omega t) .$$

Therefore:

$$v = \sqrt{v_x^2 + v_y^2} = \omega \sqrt{R^2 + r^2 - 2Rr \cos(\omega t)} .$$

For acceleration:

$$a_x = r\omega^2 \sin(\omega t) , \qquad a_y = r\omega^2 \cos(\omega t) .$$

Therefore:

$$a = \sqrt{a_x^2 + a_y^2} = r\omega^2 ,$$

Acceleration is centripetal.

Exercice 6

A wedge slides in a horizontal plane with constant acceleration.
A body moves on the inclined plane of the wedge with constant acceleration with respect to it.
At the initial instant both velocities are zero and the body is at the top of the wedge, at a height h with respect to the horizontal plane.
Determine the acceleration, velocity and trajectory of the body with respect to the plane.

We have the following scheme:

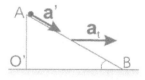

The acceleration of the body is:

$$a = a' + a_t ,$$

The components of which are:

$$a_x = a'_x + a_{tx} = a' \cos \alpha + a_t \ ,$$
$$a_y = a'_y + a_{ty} = -a' \sin \alpha \ .$$

And whose form is:

$$a = \sqrt{a_x^2 + a_y^2} = \sqrt{a'^2 + a_t^2 + 2a' a_t \cos \alpha} \ .$$

The vector a is inclined with respect to the horizontal plane by an angle:

$$\phi = \arctan \frac{a_y}{a_x} = -\frac{a' \sin \alpha}{a' \cos \alpha + a_t} \ .$$

The speed of the body is:

$$v(t) = \int_0^t a(t') \, dt' = a \, t \ .$$

The trajectory of the body is:

$$r(t) = r_0 + \int_0^t v(t') \, dt' = r_0 + \frac{1}{2} a t^2 .$$

The components of which are:

$$x(t) = \frac{1}{2} a_x t^2 = \frac{1}{2} a' t^2 \cos \alpha + \frac{1}{2} a_t t^2 \ ,$$
$$y(t) = h + \frac{1}{2} a_y t^2 = h - \frac{1}{2} a' t^2 \sin \alpha \ .$$

By eliminating the parameter, we have:

$$y = -\frac{2a' \sin \alpha}{a' \cos \alpha + a_t} x + h \ .$$

The trajectory is straight.

Exercise 7

A circular platform rotates with respect to the ground with constant angular velocity on the vertical axis.
A person runs on the platform along a circular path of radius r, in the opposite direction to the direction of rotation of the platform.
Determine the acceleration of man with respect to the ground.

With respect to the platform, the speed of the person is:

$$v' = \omega' r' = \text{costante} ,$$

And the acceleration is:

$$a' = \frac{v'^2}{r'} = \omega'^2 r' .$$

With respect to the ground, the acceleration is given by:

$$a = a' + \omega_t \times (\omega_t \times r') + 2\,\omega_t \times v' .$$

The first addend is directed radially with modulus equal to:

$$a' = \omega'^2 r'.$$

The second summand is the drag acceleration directed radially and with module equal to:

$$a_t = \omega_t^2 r' .$$

The third summand is the complementary Coriolis acceleration directly radially outward with modulus equal to:

$$a_c = 2\omega_t v' = 2\omega_t \omega' r' .$$

The acceleration will therefore be directed radially towards the center with a module equal to:

$$a = \omega'^2 r' + \omega_t^2 r' - 2\omega_t\omega' r'$$
$$= r'(\omega'^2 + \omega_t^2 - 2\omega'\omega_t)$$
$$= r'(\omega' - \omega_t)^2 .$$

Exercise 8

Consider the connecting rod-crank system shown in the figure:

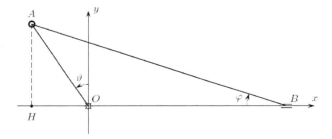

Point O is hinged and point B is constrained to slide along the x-axis.
Assume:

$$\overline{OA} = l_1, \quad \overline{AB} = l_2$$

And let the Lagrangian coordinate be the theta angle.
Determine the angular velocity of the rod AB and the velocity of point B.

The angular velocity of the first rod is:

$$\vec{\omega}_1 = \dot{\vartheta}\vec{e}_3 .$$

The speed at A is (rotary motion):

$$\vec{v}_A = \vec{\omega}_1 \wedge (A - O) = \begin{vmatrix} \vec{e}_1 & \vec{e}_2 & \vec{e}_3 \\ 0 & 0 & \dot{\vartheta} \\ -l_1 \sin\vartheta & l_1 \cos\vartheta & 0 \end{vmatrix} = -l_1\dot{\vartheta}\cos\vartheta\,\vec{e}_1 - l_1\dot{\vartheta}\sin\vartheta\,\vec{e}_2 .$$

The velocity at B is given by:

$$\vec{v}_B = \vec{v}_A + \vec{\omega}_2 \wedge (B - A) = \vec{v}_A + \vec{\omega}_2 \wedge \overrightarrow{AB}.$$

But:

$$\overrightarrow{AB} = \sqrt{l_2{}^2 - l_1{}^2 \cos^2 \vartheta}\, \vec{e}_1 - l_1 \cos \vartheta \vec{e}_2.$$

Therefore:

$$\vec{v}_B = \vec{v}_A + \begin{vmatrix} \vec{e}_1 & \vec{e}_2 & \vec{e}_3 \\ 0 & 0 & \omega_2 \\ \sqrt{l_2{}^2 - l_1{}^2 \cos^2 \vartheta} & -l_1 \cos \vartheta & 0 \end{vmatrix} =$$

$$= \left(-l_1 \dot{\vartheta} \cos \vartheta + \omega_2 l_1 \cos \vartheta\right) \vec{e}_1 + \left(-l_1 \dot{\vartheta} \sin \vartheta + \omega_2 \sqrt{l_2{}^2 - l_1{}^2 \cos^2 \vartheta}\right) \vec{e}_2.$$

Since:

$$\dot{y}_B = 0.$$

Is found:

$$\omega_2 = \frac{l_1 \dot{\vartheta} \sin \vartheta}{\sqrt{l_2{}^2 - l_1{}^2 \cos^2 \vartheta}}.$$

Exercise 9

Consider a rigid rod of length 1, whose ends are constrained to slide along the x and y axis.
Determine the trajectory described by the point P of the rod assuming that A moves with the assigned law f(t) on the x axis.

The scheme is as follows:

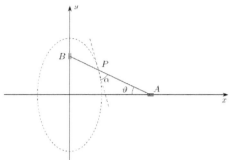

From simple geometric considerations we have:

$$x = (l - h) \cos \vartheta , \quad y = h \sin \vartheta ,$$

$$\frac{x}{l - h} = \cos \vartheta , \quad \frac{y}{h} = \cos \vartheta .$$

From which:

$$\frac{x^2}{(l - h)^2} + \frac{y^2}{h^2} = 1 ,$$

Which is the equation of an ellipse with center at the origin and semiaxes h and lh.

From Poisson's formula we have:

$$\vec{\omega} = \dot{\vartheta}\vec{k}.$$

$$x_A = f(t) = l \cos \vartheta \Rightarrow \vartheta = \arccos \frac{f(t)}{l} \Rightarrow \dot{\vartheta} = -\frac{\dot{f}(t)}{\sqrt{l^2 + f^2(t)}} .$$

$$\vec{v}_P = \vec{v}_A + \vec{\omega} \wedge \overrightarrow{AP} = \begin{vmatrix} \vec{i} & \vec{j} & \vec{k} \\ 0 & 0 & \dot{\vartheta} \\ -h\cos\vartheta & h\sin\vartheta & 0 \end{vmatrix} = \dot{f}(t)\vec{i} - h\dot{\vartheta}\sin\vartheta\vec{i} - h\dot{\vartheta}\cos\vartheta\vec{j}.$$

4

CLASSICAL MECHANICS: DYNAMICS AND STATICS

Laws of dynamics

Given m the mass of a body, i.e. the quantity of matter present in it, the momentum and angular momentum can be defined as:

$$\vec{p} = m\vec{v}$$
$$\vec{L} = \vec{r} \times \vec{p} = m\vec{v} \times \vec{r}$$

The modulus of the angular momentum is given by:

$$|L| = m\omega^2 r$$

Newton enunciated the three laws of dynamics which projected classical mechanics into a much broader perspective than kinematics could imply.
The first law of dynamics is nothing more than the acceptance of Galileo's principle of inertia.
A body not subjected to external forces or friction remains in its state of rest (if it is stationary) or of proper motion.
A particular case is given by the correct observation made at the time by Galileo in relation to a body endowed with uniform motion which remains in this state indefinitely, if it is not perturbed by external forces such as friction.
The second law of dynamics introduces the concept of force as a derivative of momentum.
In summary it is:

$$\vec{F}(\vec{r},\vec{v},t) = \frac{d\vec{p}}{dt} = m\frac{d\vec{v}}{dt} + \vec{v}\frac{dm}{dt} = ma$$

Having assumed the mass constant over time, due to a well-known conservation principle which we will explain shortly.

In the case in which the acceleration coincides with that of gravity, this force takes the name of weight-force and expresses the force of attraction of the gravitational field with respect to that mass (referring to Newton, the famous anecdote of an apple falling from a tree).

The third law of motion states that for every action there is an equal and opposite action:

$$\bar{F}_{azione} = -\bar{F}_{reazione}$$

Energy and other definitions

The mechanical energy of a system, or total energy of a system, can be defined as the sum of two energy contributions.

One takes into account the fact that the body is in motion, the other of the energy possessed by the body only due to the fact that it is potentially subject to external forces.

In the first case we speak of kinetic energy, in the second of potential energy.

These relationships exist for the mechanical energies defined as follows:

$$E_{cin} = T = \frac{1}{2}mv^2$$

$$E_{pot} = U \Rightarrow \dot{T} = -\dot{U}$$

$$E_{mecc} = W = T + U$$

Other mechanical quantities can be derived from the concept of energy.

First of all, the power given by the time derivative of the total mechanical energy:

$$P = \frac{dW}{dt} = \dot{W} = \bar{F} \cdot \vec{v}$$

The impulse is instead an expression of the difference between the quantities of momentum and will be useful in the theory of collisions.

It can be expressed as the integral of the forces:

$$\vec{S} = \Delta\vec{p} = \int \vec{F}dt$$

The work done by a force along a path is the product of the force and the displacement:

$$A = \int_{1}^{2} \vec{F} \cdot d\vec{s}$$

For the study of the dynamics of rotating systems, the concept of torque is also introduced, which is the time derivative of the angular momentum:

$$\vec{\tau} = \dot{\vec{L}} = \vec{r} \times \vec{F} = -\frac{\partial U}{\partial \vartheta}$$

The torque is also related to the derivative of the potential energy with respect to the angular coordinate, as can be clearly seen from the last equality of the equation just exposed.

Static

The condition of mechanical equilibrium on a body causes the dynamics to fall back into statics.
It occurs only if the sum of the forces (called the resultant) and the sum of the torques acting on that body are both zero.

$$\sum_{i} \vec{F}_i = 0$$

$$\sum_{i} \vec{\tau}_i = 0$$

In this case a body is precisely static, i.e. it does not translate and does not rotate as the total forces and torques cancel each other out.

A force field can be said to be conservative if it can be expressed as a gradient of a potential:

$$\bar{F}_{cons} = -\nabla U$$

From simple notations deriving from the algebra of the nabla operator, it can be seen that a conservative field is irrotational, i.e. its curl is zero:

$$\nabla \times \bar{F}_{cons} = 0$$

A fundamental consequence in a conservative field is that the work depends only on the starting and ending points, regardless of the path followed.

If the initial point coincides with the final one, as in the case of a path closed on itself, the total work is zero:

$$A = \int_{1}^{2} \bar{F} \cdot d\bar{s} = A(2) - A(1) \Rightarrow \oint \bar{F} \cdot d\bar{s} = 0$$

Conservation laws

Newton established that the gravitational force field is conservative.

Therefore, since the mechanical systems present on this planet are subjected to such forces, we have the following principles of conservation.

$$m_f = m_i$$
$$W_f = W_i$$
$$\bar{p}_f = \bar{p}_i$$
$$\bar{L}_f = \bar{L}_i$$

1) The total mass of a system is conserved
2) The total mechanical energy of a system is conserved
3) The momentum of a system is conserved
4) The angular momentum of a system is conserved

The first conservation law is actually a kind of zero principle.

The idea behind this principle is also present in many other scientific disciplines, such as chemistry or thermodynamics, about the typical phrase "nothing is created, nothing is destroyed".

60

In other words, since the mass is an expression of the quantity of matter, stating that the mass in a final state is equal to that which existed in the initial state coincides with what is explained in the sentence above.

The law of conservation of total mechanical energy states, conversely, that the total mechanical energy of a system in the final state is identical to that of the initial state.

It can therefore transform itself from kinetic energy to potential energy, or vice versa, but the sum must remain constant.

Finally, the conservation of momentum is valid for rectilinear motions while that of angular momentum is valid for rotary ones.

A remarkable result that links the different forms of energy is the so-called virial theorem.

For a single body, if potential energy can be expressed in this form

$$U = -\frac{k}{r^{\cdot n}}$$

is worth:

$$\langle m\vec{v} \cdot \vec{r} \rangle = 0 \Rightarrow \langle T \rangle = \frac{1}{2}n\langle U \rangle$$

For a multi-body system it holds instead

$$2T + U = 0 .$$

All these considerations hold when the coordinate systems are considered at rest.

Systems in motion and collision theory

Conversely, if the system itself is in motion with respect to an external observer, the above equations must be modified accordingly.

Let's say right away that this discussion is the result of studies that took place after Newton.

If we consider the kinematics and dynamics in self-motion coordinate systems, the concept of relativity with respect to reference systems and observers immediately comes into play.

Newton accepted the Galilean relativity principle and Galileo's transformations regarding the transition from one inertial system to another.
As a consequence of this, he postulated that space and time were absolute and immovable concepts, without further investigating the so-called primary essences, thus remaining loyal to the scientific method.

In a moving coordinate system, it becomes fundamental to define apparent forces to explain the effect that this motion has on the kinematic description.
The resultant of the forces for an observer placed outside the system is given by a simple subtraction:

$$\vec{F}' = \vec{F} - \vec{F}_{app}$$

If the coordinate system has translational motion, this apparent force is that with respect to the point of origin of the coordinates and is simply given by:

$$\vec{F}_{or} = -m\vec{a}_{\alpha}$$

If, on the other hand, the system has rotary motion, the apparent force depends on the angular acceleration:

$$\vec{F}_{\alpha} = -m\vec{\alpha} \times \vec{r}$$

Two particular cases of this motion occur when the reference system is the terrestrial one or in the case of uniform circular motions.
In the first case, the Coriolis force comes into play which derives from the rotation motion of the Earth on itself and is given by:

$$\vec{F}_{cor} = -2m\vec{\omega} \times \vec{v}$$

where $\vec{\omega}$ is the angular velocity of the earth.
This force is zero if v is parallel to the earth's angular velocity, i.e. if one is moving along the direction of the parallels, substantially in the east-west direction and vice versa.
This force plays a decisive role in the definition of sea currents, winds and river flows according to the hemisphere and must also be taken into due consideration for aspects of ballistic trajectory over long distances.

If instead the rotary motion is uniform circular, for the third principle of dynamics and for the condition of equilibrium, there must be an apparent force equal and opposite to the real one.
This apparent force is called centrifugal, while the real one is called centripetal.
The respective equations of these forces are:

$$\bar{F}_{cf} = m\omega^2\bar{r}_n' = -\bar{F}_{cp}$$

$$\bar{F}_{cp} = -\frac{mv^2}{r}$$

The apparent forces find an easy explanation if one passes to the tensor notation and the symbols of Levi-Civita (and this is a result of the 19th century!).
The second law of dynamics in tensor notation is expressed as follows:

$$F^\alpha = m\left\{\frac{d^2x^\alpha}{dt^2} + \Gamma^\alpha_{\beta\gamma}\frac{dx^\beta}{dt}\frac{dx^\gamma}{dt}\right\}$$

The second term in brackets is precisely due to the apparent forces.
For a multi-body system, it is convenient to define the center of mass, having the following coordinates:

$$\bar{r}_m = \frac{\sum_i m_i\bar{r}_i}{\sum_i m_i}$$

The velocity of the center of mass is given by

$$\bar{v}_{cm} = \bar{v} - \dot{\bar{R}}$$

Defined the reduced mass as the parallel sum of the masses:

$$\mu = \left(\sum_i \frac{1}{m_i}\right)^{-1}$$

63

You can separate the contributions with respect to the center of mass, i.e. separate the equations of the center of mass and with respect to the center of mass.
So the kinetic energy becomes:

$$T = \frac{1}{2}\left(\sum_i m_i\right)\dot{R}^2 + \frac{1}{2}\mu\dot{u}^2$$

In collision theory, mass, mechanical energy, momentum and angular momentum are conserved between the initial situation (before the collision) and the final one (after the collision).
The change of the relative speeds is given by the impulse:

$$\bar{S} = \Delta\bar{p} = \mu(\bar{v}_f - \bar{v}_i)$$

In a moving coordinate system, the total angular momentum is given by the following sum:

$$\bar{L}' = I\bar{\omega} + \bar{L}_n'$$

Where I is the moment of inertia defined as below:

$$I = \sum_i m_i \bar{r}_i^2$$

The kinetic energy associated with the moment of inertia is given by

$$T' = \frac{1}{2}I\omega^2$$

The value of the moment of inertia depends on the geometry of the considered body. As an example, let's give some value for some particularly useful objects:

hollow cylinder:
$$I = mR^2$$

Full cylinder:

$$I = \tfrac{1}{2} mR^2$$

Hollow sphere:

$$I = \tfrac{2}{3} mR^2$$

Full sphere:

$$I = \tfrac{2}{5} mR^2$$

Member with perpendicular axis in the center of mass:

$$I = \tfrac{1}{12} ml^2$$

Rod with axis perpendicular to the end:

$$I = \tfrac{1}{3} ml^2$$

Every rigid body has at least three principal axes, which are perpendicular to each other.
For each principal axis, the following relationship holds:

$$\frac{\partial I}{\partial \omega_x} = \frac{\partial I}{\partial \omega_y} = \frac{\partial I}{\partial \omega_z} = 0 \Rightarrow L'_n = 0$$

For time dependencies, the following equalities hold:

$$\vec{\tau}' = I \ddot{\vartheta}$$

$$\ddot{\vec{L}}' = \dot{\vec{\tau}}' - \vec{\omega} \times \vec{L}'$$

Principle of least action

An extension and generalization of classical mechanics conceived centuries after Newton's first studies is that which traces the second law of dynamics back to the principle of least action.
From this extension arose the two main descriptions of mechanics, that of Lagrange and that of Hamilton.
We note that these two approaches originally arose to describe conservative systems, but that subsequent extensions have also generalized to non-conservative systems or systems with non-perfect constraints.
The principle of least action states that, between an initial and a final instant, the motion of a physical system subjected to small perturbations causes the action to be stationary.
In mathematical terms we have:

$$\delta S = 0 \Leftrightarrow \delta \int_{a}^{b} S = 0$$

Where S is precisely the action.
The first definition of action was given by Maupertius in 1746:

$$A = \int_{t_1}^{t_2} \vec{p} \cdot \dot{\vec{q}} \, dt$$

p is the momentum and q is the generic spatial coordinate.
These coordinates are called canonical and we will shortly see their importance.
On the other hand Euler conceived an action as an effort given by the integral of the potential energy:

$$E = -\int_{t_1}^{t_2} U(q,t) \, dt$$

Finally Hamilton unified these two concepts by defining the action S as a scalar such that:

$$S = \frac{1}{2} A + E$$

Lagrangian view

At this point the variational least action problem can be traced back to different equations simply by introducing new physical quantities.

We indicate as Lagrangian the function given by the difference between kinetic energy and potential energy and we explain the dependencies of this function:

$$L(\dot{q},q,t) = T(\dot{q},q,t) - U(q,t)$$

Where q are the Lagrangian coordinates. The equations of motion deriving from the second law of dynamics and from the variational principle lead to these new equations, called Euler-Lagrange:

$$\frac{d}{dt}\frac{\partial L}{\partial \dot{q}_i} = \frac{\partial L}{\partial q_i}$$

The canonical momentum in this new notation is given by:

$$p_i = \frac{\partial L}{\partial \dot{q}_i}$$

The Lagrangian notation has two great advantages over the classical treatment given by Newton. Firstly, in the case of constrained systems it is possible to obtain the equations of motion without having to take into account the constrained reactions. This advantage is powerful in terms of calculation speed and ease of use. As an example, we report the generic Lagrangian in spherical three-dimensional coordinates:

$$L = \frac{m}{2}(\dot{r}^2 + r^2\dot{\vartheta}^2 + r^2\sin^2(\vartheta)\dot{\varphi}) - U(r,\vartheta,\varphi)$$

If we wanted to pass from a description of a material point not subject to constraints to one subject to constraints and placed at a distance d, it is enough to take this Lagrangian, setting r=d, and calculate the time-dependent functions of the other two coordinates.

The second advantage derives from the consideration that if the Lagrangian does not depend on a certain coordinate, then this relation exists:

$$\frac{d}{dt}\left(\frac{\partial L}{\partial \dot{q}^{\lambda}} \right) = 0$$

In that case

$$\frac{\partial L}{\partial \dot{q}^{\lambda}}$$

is called constant of motion or first integral, while

$$q^{\lambda}$$

is called the cyclic coordinate. This consideration, generalized in 1915 by Noether with the homonymous theorem, leads to the consequence that a conservation law corresponds to every symmetry of the Lagrangian. In particular, the law of conservation of energy corresponds to a temporal symmetry of the Lagrangian, to a translational symmetry that of the quantity of motion and to a rotational symmetry that of the angular momentum. The Lagrangian interpretation was completed in 1788.

Hamiltonian view

A subsequent extension of the Lagrangian vision was given by Hamilton in 1833. The point of connection between the two visions is given by the fact that the solutions of the Euler-Lagrange equations are the stationary points of the action, i.e. the following result of variational calculus is valid:

$$\delta \int_{a}^{b} L(q,\dot{q},t)dt = 0$$

This result is known as Hamilton's variational principle.
With this principle it is possible to give a Lagrangian formulation to the equation of geodesics.
If we apply the Legendre transform to the Lagrangian, we obtain a new function called Hamiltonian:

$$H = \sum_i \dot{q}_i p_i - L$$

This function has some notable properties:
a) If the Lagrangian does not depend on time, the Hamiltonian is a time constant.
b) If the Lagrangian is given by the difference between kinetic energy and potential energy, the Hamiltonian is their sum, ie it is the total mechanical energy.
The equations of motion are expressed by Hamilton's equations:

$$\frac{\partial H}{\partial p_i} = \dot{q}^i$$

$$-\frac{\partial H}{\partial q^i} = \dot{p}_i$$

Similar considerations to those made on the Lagrangian apply as regards the constants of motion.
The Hamiltonian for a free particle is given by:

$$H = \frac{p^2}{2m}$$

While for a one-dimensional harmonic oscillator it is given by:

$$H = \frac{p^2}{2m} + \frac{1}{2} m \omega^2 x^2$$

In this case, if the canonical coordinates of a harmonic oscillator are considered, the Hamiltonian is very simplified in the form:

$$\vartheta = \arctan\left(\frac{-p}{m \omega x} \right)$$

$$I = \frac{p^2}{2m\omega} + \frac{1}{2} m \omega x^2$$

$$H = \omega I$$

In Hamiltonian notation, the canonical variables commute:

$$\{q_i, q_j\} = 0$$
$$\{p_i, p_j\} = 0$$
$$\{q_i, p_j\} = \delta_{ij}$$

Having defined the Poisson brackets as follows:

$$\{A, B\} = \sum_i \left[\frac{\partial A}{\partial q_i} \frac{\partial B}{\partial p_i} - \frac{\partial A}{\partial p_i} \frac{\partial B}{\partial q_i} \right]$$

The Hamiltonian formalism is particularly effective in defining the principles of conservation and the resulting continuity equations.
The divergence of a generic vector in the phase space described by the Hamiltonian vector field is given by:

$$\nabla \cdot \vec{v} = \sum_i \left(\frac{\partial}{\partial q_i} \frac{\partial H}{\partial p_i} - \frac{\partial}{\partial p_i} \frac{\partial H}{\partial q_i} \right)$$

The generic continuity equation, which in classical mechanics is expressed in the following form:

$$\partial_t \rho + \nabla \cdot (\rho \vec{v}) = 0$$

With Hamilton's formalism it becomes simply:

$$\frac{\partial \rho}{\partial t} + \{\rho, H\} = 0$$

Liouville's theorem imposes the following conditions (with C a suitable constant):

$$\frac{d\rho}{dt} = 0 \leftrightarrow \int p\,dq = C$$

In this case the generic scalar is conserved.

Therefore a quantity is conserved if it commutes with the Hamiltonian.

From this consideration it can be deduced that the passage from one Hamiltonian form to another is possible if four generating functions are identified.

Given therefore any mechanical system, one can always go back to a known system with a simple change of coordinates originating from the explicitation of the generating functions.

A final generalization of classical mechanics is given by the Hamilton-Jacobi equation:

$$H + \frac{\partial S}{\partial t} = 0$$

Where S is the action.

This equation derives from those of Hamilton when S is considered as the generating function of a canonical transformation of the Hamiltonian.

In this notation, the conjugate moments are expressed as follows:

$$p_k = \frac{\partial S}{\partial q_k}$$

$$\bar{p} = \frac{\partial L}{\partial \dot{\bar{q}}}$$

While the constants of motion are simply given by the derivative of the action with respect to the constants of the above canonical transformation:

$$\beta_k = \frac{\partial S}{\partial \alpha_k}$$

In all the views reported up to now, it is easy to study the cases of small perturbations around the equilibrium by applying the well-known linearization procedures and the principle of superposition of effects.

High school exercises

Exercise 1

A sample kilogram is accelerated by:

$$\vec{F_1} = 3.0\,\vec{i} + 4.0\,\vec{j} \quad \text{ed} \quad \vec{F}_2 = -2.0\,\vec{i} - 6.0\,\vec{j}.$$

Determine the net net force, the magnitude and direction of both the force and the acceleration.

The resultant force is the vector sum of the two given forces:

$$\vec{R} = (3.0 - 2.0)\,\vec{i} + (4.0 - 6.0)\,\vec{j} = 1.0\,\vec{i} - 2.0\,\vec{j}$$

The intensity of the force is:

$$R = \sqrt{(1.0)^2 + (-2.0)^2} = 2.2\,N$$

While the direction is given by the angle calculated as follows:

$$\alpha = \arctan\left(\frac{-2.0}{1.0}\right) = 116.6°$$

The acceleration will have the same direction as the force and intensity equal to:

$$a = \frac{F}{m} = \frac{2.2\,N}{1.0\,kg} = 2.2\,\frac{m}{s^2}$$

Exercise 2

An object has a mass of 75 kg on Earth.
Knowing that on Mars the acceleration due to gravity is equal to 38.73% of that of the earth and in space this acceleration is zero, calculate the weight and mass of the object on Earth, on Mars and in space.

Mass is an intrinsic quantity of the body and therefore will always be equal to 75 kg.
On Earth and Mars the weight is given by:

$$P_{Terra} = 75\,kg \cdot 9.8\,\frac{m}{s^2} = 735\,N$$

$$P_{Marte} = 75\,kg \cdot 3.8\,\frac{m}{s^2} = 285\,N$$

While in space it is zero because the acceleration due to gravity is zero.

Exercise 3

An object of mass 500 kg is constantly accelerated from rest to a speed of 1'600 km/h in 1.8 seconds.
Find the amount of average force needed.

The speed in the international system is given by:

$$\frac{1600}{3.6} = \bar{444.4}\,m/s$$

The motion is uniformly accelerated so:

$$a = \frac{\Delta v}{\Delta t} = \frac{(444.4 - 0)\,\frac{m}{s}}{1.8\,s} = 247\,\frac{m}{s^2}$$

The average force is therefore:

$$F = ma = 500\,kg \cdot 247\,\frac{m}{s^2} = 123500\,N = 1.2 \cdot 10^5\,N$$

Exercise 4

An object of mass 8.5 kg slides without friction on a plane inclined by 30° and is kept in balance by means of a rope fixed to a wall.
Find the tension T in the rope and the normal force N acting on the object.

Find the acceleration of the object if the rope is severed.

The problem can be schematized graphically:

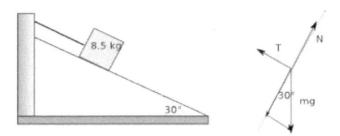

We decompose the weight force into the two direct components such as the inclined plane and perpendicular to it (ie as T and as N).
Recalling the properties of right triangles of 30° and 60° (or indifferently applying the trigonometric relations in right triangles) we have:

$$(mg) = P_{parallelo} \quad = \quad mg \cdot \frac{1}{2} = \frac{8.5\,kg \cdot 9.8\,\frac{m}{s^2}}{2} = 42\,N$$

$$(mg) = P_{perp} \quad = \quad mg \cdot \frac{\sqrt{3}}{2} = \frac{8.5\,kg \cdot 9.8\,\frac{m}{s^2} \cdot \sqrt{3}}{2} = 72\,N$$

Since the object is in equilibrium, it must necessarily be:

$$T \quad = \quad -P_{parallelo} = -42\,N$$
$$N \quad = \quad -P_{perp} = -72\,N$$

If the rope is sheared, the only acceleration component to act on the object is the parallel one and therefore:

$$a = \frac{P_{parallelo}}{m} = \frac{-42}{8.5\,kg} = 4.9\,\frac{m}{s^2}$$

Exercise 5

A sphere of mass 0.3 grams is suspended by a string. A constant horizontal force causes it to move so that the string makes an angle of 37° with the vertical.
Find the magnitude of the horizontal force and the tension in the thread.

The problem can be schematized graphically:

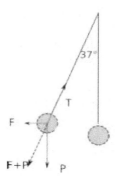

A weight force equal to:

$$P = mg = 3.0 \cdot 10^{-4} \, kg \cdot 9.8 \frac{m}{s^2} = 2.9 \cdot 10^{-3} \, N$$

For the balance of forces we have:

$$\vec{F} + \vec{P} = -\vec{T}$$

On the other hand, from the trigonometry formulas applied to right triangles we have that:

$$\frac{\vec{F}}{\vec{P}} = \tan 37°$$

And so the horizontal force has magnitude:

$$\vec{F} = 2.9 \cdot 10^{-3} \, N \cdot \tan 37° = 2.2 \cdot 10^{-3} \, N$$

To find the tension in the thread we apply the Pythagorean theorem to the magnitude of the forces:

$$\vec{F} + \vec{P} = -\vec{T} = \sqrt{(2.9 \cdot 10^{-3}\,N)^2 + (2.2 \cdot 10^{-3}\,N)^2} = -3.6 \cdot 10^{-3}\,N$$

Exercise 6

An 85 kg object is lowered to the ground from a height of 10 meters using a pulley with a 65 kg counterweight.
Starting from rest, calculate the speed with which the object hits the ground.

Calling T the counterweight, the balance of forces on the counterweight dictates that:

$$T - mg = ma$$

Where m is the mass of the counterweight.
On the other hand, on the object, called M its mass, there is a relation of the type:

$$T - Mg = -Ma$$

Since the acceleration with which the object descends is the same as that with which the counterweight rises.
Combining these two relations, we get:

$$a = \frac{M - m}{M + m}g = \frac{85 - 65}{85 + 65} \cdot 9.8\,\frac{m}{s^2} = 1.3\,\frac{m}{s^2}$$

The motion will be uniformly accelerated and the final speed will be given by:

$$v = \sqrt{2ha} = \sqrt{2 \cdot 10.0\,m \cdot 1.3\,\frac{m}{s^2}} = 5.1\,\frac{m}{s}$$

Exercise 7

An object of mass 3.7 kg is placed on a plane inclined by 30° and is connected by a rope which, through a pulley, joins it to another object of mass 2.3 kg suspended vertically.
Find the acceleration of each block, the directions of those accelerations, and the tension in the string.

The problem can be schematized graphically:

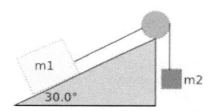

Recalling the properties of right triangles of 30° and 60° on the first block, the balance of forces is as follows:

$$T - m_1 g \sin 30.0° = m_1 a$$

While on the second block we have:

$$T - m_2 g = -m_2 a$$

Putting it to the system, we find:

$$T = m_1 a + \frac{1}{2} m_1 g$$

$$m_1 a + \frac{1}{2} m_1 g - m_2 g = -m_2 a$$

And then:

$$a = \frac{m_2 - \frac{1}{2} m_1}{m_1 + m_2} g = \frac{2.30\,kg - 1.85\,kg}{6.0\,kg} \cdot 9.8 \frac{m}{s^2} = 0.735 \frac{m}{s^2}$$

This acceleration is the same for the two blocks.

For the first block it means that the acceleration is directed upwards of the inclined plane, while for the second block that this acceleration is directed downwards.

The tension in the string is:

$$T = m_2 (g - a) = 2.30 \, kg \cdot (9.8 - 0,735) \, \frac{m}{s^2} = 20.8 \, N$$

Exercise 8

Given two objects, the first has a kinetic energy equal to half that of the second.

The mass of the second object is half that of the first.

If the first object increases its speed by 1 m/s then it equals the kinetic energy of the second object.

Find the initial velocities of objects.

Calling p the first object and f the second object, we have:

$$m_p = 2m_f$$
$$K_p = \frac{1}{2}K_f$$

Or:

$$\frac{1}{2}m_p v_{i,p}^2 = \frac{1}{4}m_f v_f^2$$

But substituting the masses:

$$m_f v_{i,p}^2 = \frac{1}{4}m_f v_f^2$$

And then:

$$v_f = 2v_{i,p}$$

On the other hand we know that:

$$\frac{1}{2}m_p v_{f,p}^2 = \frac{1}{2}m_f v_f^2$$

$$v_{f,p} = v_{i,p} + 1,$$

Or:

$$(v_{i,p} + 1)^2 = \frac{1}{2}v_f^2$$

Recalling what was said earlier:

$$(v_{i,p} + 1)^2 = \frac{4v_{i,p}^2}{2}$$

From which:

$$v_{i,p} = 2.4\,\frac{m}{s}$$

$$v_{f,p} = 4.8\,\frac{m}{s}$$

Exercise 9

An object of mass 102 kg moves in a straight line with a speed of 53 m/s. We want to stop it with a deceleration equal to 2 meters per second frame. Determine the amount of force required, the distance traveled during the slowdown and the work done by the slowing force.

The decelerating force is:

$$F = ma = 102\,kg \cdot 2.0\,\frac{m}{s^2} = 204\,N$$

The kinetic energy of the body before coming to rest is given by:

$$K = \frac{1}{2}mv^2 = 0.5 \cdot 102\,kg \cdot \left(53\,\frac{m}{s}\right)^2 = 143259\,J$$

Since the body will eventually be stationary, its final kinetic energy will be zero.
Therefore, for the conservation of energy, the work will be:

$$W = K_{fin} - K_{ini} = -143259\,J$$

The distance traveled is:

$$s = \frac{W}{F} = \frac{143259\,J}{204\,N} = 702\,m$$

Exercise 10

An object is pushed a space given by:

$$(15\,m)\,\vec{i} - (12\,m)\,\vec{j}$$

From the following strength:

$$\vec{F} = (210\,N)\,\vec{i} - (150\,N)\,\vec{j}.$$

Finding work developed by that force.

The problem can be schematized graphically as:

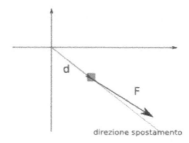

direzione spostamento

The displacement has an angle with respect to the x axis given by:

$$\beta_{spos} = \tan \frac{12}{15} = 38.7°$$

While the force is directed:

$$\beta_F = \tan \frac{150}{210} = 35.5°$$

The angle between the two vectors is:

$$\alpha = 38.7° - 35.5° = 3.2°$$

The magnitude of the displacement and the force are:

$$s = \sqrt{144 + 225} = 19.2\,m$$
$$F = \sqrt{210^2 + 150^2} = 258\,N$$

So the work done is:

$$W = Fs\cos\alpha = 258\,N \cdot 19.2\,m \cdot \cos 3.2° = 4946\,J$$

Exercise 11

At your place:

$$F_1 = 5.00\,N,\ F_2 = 9.00\,N,\ F_3 = 3.00\,N$$

Arranged as shown:

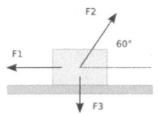

The object is known to move 3 meters to the left.
Calculate the net work during the move.

The resultant of the three forces along the horizontal axis (we consider the displacement to the left positive) is given by:

$$
\begin{aligned}
F_1 &= 5.00\,N \\
F_2 &= \frac{9.00}{2} = -4.50\,N \\
F_3 &= 0\,N
\end{aligned}
$$

For the second force we have applied the well-known properties of right triangles of 30° and 60°.
The net result will therefore be:

$$F = 0.50\,N$$

The work done is:

$$W = Fs = 0.50\,N \cdot 3.00\,m = 1.50\,J$$

Exercise 12

An object of mass 25 kg is placed on a plane inclined by 25° and is subjected to a force of 209 N parallel to the inclined plane in an ascending direction.
Find the work done by the parallel force, the gravitational force and the normal force for a displacement of 1.5 meters.
Then find the total work.

The problem can be schematized graphically as:

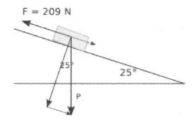

The weight force and the parallel force will be:

$$P = mg = 25.0\,kg \cdot 9.8\,\tfrac{m}{s^2} = 245\,N$$
$$P_{par} = 245 \cdot \sin 25° = 104\,N$$

Having used a trigonometry theorem for right triangles. Similarly the normal force will be:

$$P_{per} = 245 \cdot \cos 25° = 222\,N$$

The work of the force is given by:

$$W_F = 209\,N \cdot 1.5\,m = 314\,J$$

While the resistant one of the weight force:

$$W_P = 245\,N \cdot 1.5 \cdot \cos 115° = -155\,J$$

Since the work of the normal force is zero (since it is perpendicular to the displacement), the resulting work will be:

$$W_{tot} = 314 - 155 = 159\,N$$

Exercise 13

A force of 5 N acts on an object of mass 15 kg which is initially at rest.

Calculate the work done by the force in 3 seconds and the power developed.

The acceleration imparted to the body is:

$$a = \frac{F}{m} = \frac{5.0\,N}{15\,kg} = 0.33\,\frac{m}{s^2}$$

Starting from rest, it will have a final speed given by:

$$v = at = 0.33\,\frac{m}{s^2} \cdot 3\,s = 1.0\,\frac{m}{s}$$

The work will therefore be equal to the final kinetic energy:

$$L = \Delta K = \frac{1}{2} \cdot 15.0\,kg \cdot 1^2\,\frac{m^2}{s^2} = 7.5\,J$$

While the power:

$$P = Fv = 5.0\,N \cdot 1\,\frac{m}{s} = 5.0\,W$$

Exercise 14

Determine the constant of a spring that stores 25 J of elastic potential energy when compressed by 7.5 cm from its equilibrium position.

The potential energy of a spring is:

$$\Delta U = \frac{1}{2}k\left(x_f^2 - x_i^2\right)$$

Assuming zero as the equilibrium position:

$$\Delta U = \frac{1}{2}kx^2$$

And then its spring constant:

$$k = \frac{2\Delta U}{x^2} = \frac{2 \cdot 25\,J}{(0.075\,m)^2} = 8889\,\frac{N}{m}$$

Exercise 15

An object of mass 2 kg falls from a height of 10 meters.
Calculate the initial potential energy, the kinetic energy at a height of 1.5 meters and the speed of the fall at that instant.

The potential energy is given by:

$$U = mgh$$

At 10 meters high it will be:

$$U_{10} = 2.0\,kg \cdot 9.8\,\frac{m}{s^2} \cdot 10\,m = 196\,J$$

At 1.5 meters high it will be:

$$U_{1.5} = 2.0\,kg \cdot 9.8\,\frac{m}{s^2} \cdot 1.5\,m = 29.4\,J$$

For the conservation of total mechanical energy, the potential energy difference will be given by the kinetic energy:

$$K = \Delta U = (196 - 29.4)\,J = 166.6\,J$$

So the final speed will be:

$$v = \sqrt{\frac{2K}{m}} = \sqrt{\frac{2 \cdot 166.6\,J}{2.0\,kg}} = 12.9\,\frac{m}{s}$$

Exercise 16

Given the following path:

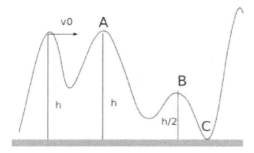

Find the velocity at point A, point B, and point C.
Determine the maximum height that can be reached.

At point A the speed will be equal to the initial speed, as it is at the same height with respect to the initial position.
At point B, part of the potential energy will have been transformed into kinetic energy.
In particular you will have:

$$\Delta U = mgh - \frac{1}{2}mgh = \frac{1}{2}mgh$$

By decomposing along the two axes:

$$v_{By} = \sqrt{2\frac{h}{2}g} = \sqrt{hg}$$

And then:

$$v_B = \sqrt{v_0^2 + hg}$$

In C all the potential energy will have become kinetic energy and therefore the speed will be:

$$v_C = \sqrt{v_0^2 + 2hg}$$

Having the component along y reached the maximum value (while the one in x always remains equal to the initial speed).

At this point, when all the kinetic energy in C has been transformed into potential energy, we will have the maximum height:

$$\frac{1}{2}mv_C^2 = mgh_{max}$$

Which will be equal to:

$$h_{max} = \frac{v_C^2}{2g} = \frac{v_0^2 + 2hg}{2g} = \frac{v_0^2}{2g} + h$$

Exercise 17

An object of mass 8 kg compresses a spring of 10 cm. Find the spring constant.

The object is pushed another 30cm and then released.

Find the potential energy of the spring before release.

If the object is thrown upwards, determine the maximum height that can be reached.

The weight force of the object is:

$$P = mg = 8\,kg \cdot 9.8\,\frac{m}{s^2} = 78.4\,N$$

Knowing that the force on a spring is:

$$F = -k\Delta x,$$

The spring constant is given by:

$$k = \frac{78.4\,N}{0,10\,m} = 784\,\frac{N}{m}$$

The potential energy of a spring is:

$$U = \tfrac{1}{2}kx^2$$

Since the total displacement is 40 cm, we have:

$$U = \frac{1}{2} \cdot 784\,\frac{N}{m} \cdot (0.40)^2\,m^2 = 62.7\,J$$

The available energy will all transform into kinetic energy after the release therefore the speed of the object will be:

$$v = \sqrt{\frac{2K}{m}} = \sqrt{\frac{2 \cdot 62.7\,J}{8\,kg}} = 4\,\frac{m}{s}$$

Being a uniformly decelerated motion, the maximum height will be:

$$h = \frac{v^2}{2g} = \frac{16\,\frac{m^2}{s^2}}{2 \cdot 9.8\,\frac{m}{s^2}} = 0.82\,m$$

Exercise 18

A spring can be compressed by 2 cm by a force of 270 N. An object of mass 12 kg is placed on top of a plane inclined by 30° and is initially at rest.
The object is released and stops after compressing the spring 5.5cm.
Find how far it has moved down the incline and the speed of the object when it touches the spring.

The spring constant is:

$$k = \frac{F}{\Delta x} = \frac{270 \, N}{2 \, cm} = 135 \, \frac{N}{cm}$$

At the initial instant we will have:

$$U = mgh$$
$$K = 0.$$

While when the object touches the spring:

$$U = mgh'$$
$$K = \tfrac{1}{2}mv^2.$$

The potential energy of the spring after compression is:

$$U = \tfrac{1}{2}k\Delta x^2$$

Therefore:

$$\frac{1}{2}k\Delta x^2 = mg\,(h - h')$$

From which:

$$h - h' = \frac{k}{2gm}\Delta x^2 = \frac{135 \, \frac{N}{cm}}{2 \cdot 9.8 \, \frac{m}{s^2} \cdot 12 \, kg}\,(5.5)^2 \, cm^2 = 17.4 \, cm$$

Knowing that the plane is inclined by 30°, we find:

$$\Delta l = 35 \, cm$$

The speed will therefore be given by:

$$\sqrt{2g\,(h - h')} = \sqrt{2 \cdot 9.8 \, \frac{m}{s^2} \cdot 0.174 \, m} = 1.85 \, \frac{m}{s}$$

Exercise 19

In a simple pendulum of length L, the weight at the lower end has a given velocity when the pendulum makes a given angle with the vertical.
Find an equation expressing the speed of the weight in the lowest position.
Find the minimum speed value for the pendulum to swing 90° vertically.

The problem can be schematized graphically as:

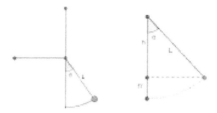

Where is it:

$$h' = L - h = L - L\cos\theta_0 = L\left(1 - \cos\theta_0\right)$$

The speed increase is given by:

$$v^2 = v_0^2 + 2gh' = v_0^2 + 2gL\left(1 - \cos\theta_0\right)$$

The minimum speed for which the 90° oscillation takes place is that in which the kinetic energy in the lowest point is totally transformed into potential energy in the horizontal position, i.e.:

$$mgL = \frac{1}{2}m\left[v_0^2 + 2gL\left(1 - \cos\theta_0\right)\right]$$

From which:

$$v_0 = \sqrt{2gL\cos\theta_0}$$

90

Exercise 20

An object weighing 267 N falls down a 20° inclined plane which has a length of 6.1 meters.
The coefficient of dynamic friction is 0.1
Find the work done on the object, the energy dissipated by the frictional force, and the final velocity if the initial velocity is 0.457 m/s.

The height of the inclined plane is:

$$h = 6.1\,m \times \quad \sin 20° \;=\; 2.1\,m.$$

Having used a well-known trigonometry theorem.
The final potential energy is zero, so the work is equal to the initial potential energy:

$$W_{peso} = \Delta U = Ph = 267\,N \times 2.1\,m = 557\,J$$

The energy dissipated by the frictional force along the 6.1 meters of the plane depends only on the perpendicular component of the force:

$$W_{att} = -\mu P_{perp} l = 0.1 \times 267\,N \times \cos 20° \times 6.1\,m = 153\,J$$

So the kinetic energy will be:

$$\Delta K = 557 - 153 = 404\,J$$

Knowing that:

$$\frac{1}{2}m\left(v_f^2 - v_i^2\right) = 404\,J$$

Find the final speed:

$$v_f = \sqrt{\frac{2 \times 404J \times 9.8\,\frac{m}{s^2}}{267\,N}} + 0.457\,\frac{m}{s^2} = 5.45\,\frac{m}{s}$$

Exercise 21

An object glides along a path with two equal curved vertical strokes and a flat central stroke of length L.

The two curved parts have no friction, while the flat part has a dynamic coefficient of friction of 0.2.

The object is allowed to fall from a height equal to half L.

Find where the object stops.

The initial potential energy is:

$$U = mgh = 0.5mgL$$

And it is converted into kinetic energy on the way down.
On a flat section, friction does work equal to:

$$W = -F_{att}L = -0.20mgL$$

So the kinetic energy at the end of the flat section will be:

$$K = mgL \left(0.5 - 0.20\right) = 0.30mgL$$

This kinetic energy will be transformed into potential energy by making the ball rise up to the height of:

$$0.30mgL = mgh'$$

$$h' = 0.30L = 0.60h$$

The particle will descend again and on the second passage in the flat section it will lose more energy equal to:

$$W = -F_{att}L = -0.20mgL$$

It will ascend and descend to a stop midway up the flat section in the third pass, with only half the kinetic energy left to overcome the full force of friction.

University-level exercises

Exercise 1

A force of components:

$$F_x = 4, \ F_y = 2, \ F_z = 0$$

It is applied to the coordinate point:

$$x_p = 1, \ y_p = 1, \ z_p = 0$$

Find the moment of force about the origin.

Having said P the point, the moment of force is:

$$\tau_0 \ = \ r \times F \ = \ OP \times F \ .$$

Therefore:

$$\begin{vmatrix} \hat{\imath} & \hat{\jmath} & \hat{k} \\ r_x & r_y & r_z \\ F_x & F_y & F_z \end{vmatrix} \ = \ \begin{vmatrix} \hat{\imath} & \hat{\jmath} & \hat{k} \\ 1 & 1 & 0 \\ 4 & 2 & 0 \end{vmatrix} \ = \ 0\,\hat{\imath} + 0\,\hat{\jmath} - 2\,\hat{k} \ = \ -2\,\hat{k} \ .$$

Exercise 2

A man of weight P balances a suspended platform AB by pulling F on one end of a rope.
If the system is the one in the figure:

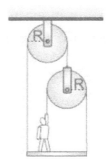

With the two pulleys having the same radius R and the platform being 3R long, determine the magnitude of the force F.
What if the two pulleys have different radius?

Balancing on pulley 2 requires:

$$T_1 = F + T_2 = 2F.$$

The balance on the platform instead:

$$\sum_i F_{iy} = 0,$$
$$\sum_i T_i = 0.$$

Or:

$$P = F + T_1 + T_2 \quad \Rightarrow \quad P = 4F \quad \Rightarrow \quad F = P/4.$$
$$2RT_2 = RT_1 \quad \Rightarrow \quad T_1 = 2T_2.$$

If the two pulleys had different radii, we would have:

$$T_1 = 2T_2 = 2F, \qquad F = P/4.$$

$$T_1 \left(2R_1 - R_2\right) = 2\,T_2 R_2 \, ,$$

$$T_1 = 2\,T_2 \, \frac{R_2}{2R_1 - R_2} \, .$$

The system is evidently incompatible and there would be no equilibrium.

Exercise 3

A ball of negligible size can slide frictionlessly on a semicircular concave surface of radius R.
The ball is made to start from one end of the surface with zero initial speed.
Determine the equation of motion of the ball.
Determine the trend of the angular velocity as a function of the angle.
Determine the reaction N as a function of the angle.

For the balance of forces it must be:

$$mg + N = ma \, .$$

Where the acceleration is given by:

$$a = a_T \, \hat{u}_T + a_N \, \hat{u}_N \, .$$

And so the equation becomes:

$$
\begin{aligned}
x : & \qquad mg \cos \theta & = ma_T \, , \\
y : & \qquad N - mg \sin \theta & = ma_N \, .
\end{aligned}
$$

But:

$$a_T = dv/dt$$

$$v = \omega R.$$

Therefore:

$$R \frac{d\omega}{dt} = g \cos \theta .$$

$$R\omega \, d\omega = g \cos \theta \, d\theta .$$

By integrating we have:

$$R \int_0^\omega \omega' \, d\omega' = g \int_0^\theta \cos \theta' \, d\theta' .$$

So:

$$\omega(\theta) = \pm \sqrt{\frac{2g}{R}} \sin \theta .$$

Also being:

$$a_N = \omega^2 R .$$

We have:

$$N = mg \sin \theta + m\omega^2 R = 3 \, mg \sin \theta .$$

Exercise 4

A point on a surface is subject to:

$$F_x = 0 , \qquad F_y = kx .$$

Calculate the work for a shift from (-1,-1) to (1,1).

To calculate the work it is necessary to study the trajectory. Following this first trajectory:

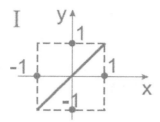

We have that the work is:

$$W \ = \ \int_{P_1}^{P_2} F \cdot \mathrm{d}r \ = \ \int_{P_1}^{P_2} (F_x \, \mathrm{d}x + F_y \, \mathrm{d}y) \ = \ \int_{x_1}^{x_2} F_x \, \mathrm{d}x \ + \ \int_{y_1}^{y_2} F_y \, \mathrm{d}y \ .$$

Therefore:

$$W \ = \ \int_{-1}^{+1} F_y \, \mathrm{d}y \ = \ k \int_{-1}^{+1} x \, \mathrm{d}y \ .$$

Or:

$$W_I \ = \ k \int_{-1}^{+1} y \, \mathrm{d}y \ = \ k \frac{y^2}{2} \bigg|_{-1}^{+1} \ = \ 0 \ .$$

Following instead:

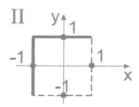

We have:

$$W_{II} = \int_{P_1}^{P_3} \mathbf{F} \cdot d\mathbf{r} + \int_{P_3}^{P_2} \mathbf{F} \cdot d\mathbf{r} .$$

$$W_{II} = k \int_{-1}^{+1} (-1) \, dy = -ky \Big|_{-1}^{+1} = -2k \neq 0.$$

The force field is not conservative.

Exercise 5

A cart of mass m slides along the following track:

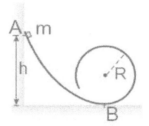

Determine the minimum height for the carriage to travel the entire circular path.
Determine the reaction N as a function of the central angle.

For the balance of forces on the trolley we have:

$$mg + N = ma .$$

And then:

$$mg + N_M = mv_M^2 / R ,$$

It must be that:

98

$$v_M^2 \geq gR .$$

Or:

$$mgh = mv_M^2/2 + 2mgR ,$$

$$h = v_M^2/2g + 2R .$$

And then:

$$h_{min} = R/2 + 2R = 2.5\,R .$$

Rewriting the equation of motion by making alpha explicit, we have:

$$N - mg\cos\alpha = mv_P^2/R .$$

From the conservation of total mechanical energy:

$$mgh = mv_P^2/2 + mgR\,(1 - \cos\alpha) .$$

Finding the speed and substituting, we have:

$$N = mg\,(2h/R + 3\cos\alpha - 2) .$$

Exercise 6

A material point of mass m moves on a frictionless horizontal plane along a circular path of radius R, with constant angular velocity.
A rope, tied at one end to the point, passes through a hole in the horizontal plane. A force F is applied to the other end of the rope.
Determine the modulus of F.
If the modulus is slowly increased to another value, how do the angular velocity and radius change?
Calculate the work required to increase the force by that value.

By the law of dynamics we have:

$$F' = ma; \qquad F = F' = m\omega^2 R.$$

Passing from a state 1 to a state 2, for the conservation of angular momentum we have:

$$L_1 = L_2.$$

$$L_1 = mR_1v_1 = mR_1^2\omega_1,$$
$$L_2 = mR_2v_1 = mR_2^2\omega_2,$$

Therefore:

$$\frac{\omega_2}{\omega_1} = \frac{R_1^2}{R_2^2}.$$

From which:

$$\frac{F_1}{F_2} = \frac{m\omega_1^2 R_1}{m\omega_2^2 R_2} = \frac{R_2^3}{R_1^3}.$$

So as the force increases, the radius decreases and the angular velocity increases.

According to the principle of conservation of energy, work equals the change in kinetic energy:

$$W = E_{k,2} - E_{k,1} = \frac{1}{2}mv_2^2 - \frac{1}{2}mv_1^2$$
$$= \frac{1}{2}m\left(\omega_2^2 R_2^2 - \omega_1^2 R_1^2\right) = \frac{1}{2}m\omega_1^2 R_1^2 \left(\frac{R_1^2}{R_2^2} - 1\right).$$

Exercise 7

A body of mass 100 kg is placed on a balance with negligible mass which is fixed to a platform of 200 kg.
The platform slides without friction along a 30° inclined plane.
What mass value will be read on the balance dial?

For the balance of forces, the equation of motion is:

$$(M + m)\, g + N = (M + m)\, a_t$$

$$a_t = g \sin \theta .$$

For an inertial observer we have:

$$mg + F_e + T = ma_t .$$

And then:

$$F_e = mg\,(1 - \sin^2 \theta) = mg \cos^2 \theta .$$

That is 750 N.
For a non-inertial observer we have:

$$mg - ma_t + F_e + T = ma' = 0$$

And then:

$$F_e = mg \cos^2 \theta .$$

So the same result as before.

Exercise 8

A body of mass m rests on a plane inclined by an angle theta.
The body, initially at rest at a height h, is left free to move.

There is a coefficient of kinetic friction.
With what speed does the body arrive at the bottom of the inclined plane?
What distance d will it travel on the plane before coming to rest?

The force of gravity, the normal reaction and the force of friction act on the body:

$$P = mg;$$
$$N = mg \cos \theta;$$
$$F_a = \mu N = \mu mg \cos \theta.$$

The reaction N does no work as it is orthogonal to the displacement, so the change in kinetic energy is:

$$\Delta E_k = m v_1^2/2 = W_{peso} + W_{attr} ,$$

The weight force is conservative therefore:

$$W_{peso} = - \Delta E_p = mgh = mgs \sin \theta .$$

The work of the frictional force is:

$$W_{attr} = \int_{-s}^{0} F_a \cdot dx = \int_{-s}^{0} - F_a \, dx = (- \mu mg \cos \theta) s .$$

And then:

$$\Delta E_k = m v_1^2/2 = mgs \sin \theta - \mu mgs \cos \theta = mgs (\sin \theta - \mu \cos \theta)$$

From which:

$$v_1 = \sqrt{2gs (\sin \theta - \mu \cos \theta)} .$$

In order for the body to move it must be:

$$\mu > \tan \theta .$$

On the horizontal plane there is only:

$$F'_a = \mu m g ,$$

For the conservation of energy it will be:

$$\Delta E_k = 0 - m v_1^2 / 2 = W_{attr} .$$

But:

$$W_{attr} = \int_0^d F'_a \cdot dy = (-\mu m g) d ,$$

So:

$$d = \frac{v_1^2}{2\mu g} = \frac{s}{\mu} (\sin\theta - \mu \cos\theta) .$$

Exercise 9

Given:

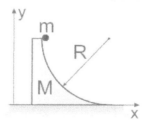

Determine the velocity of the body and profile along the x-axis once the body has slid to the base of the profile.

Determine the speed of the center of mass of the system as a function of the angle which identifies the position of the body on the profile.

For the horizontal direction we have:

$$\left(\sum F_{\text{ext}} \right)_x = 0 .$$

Hence the horizontal component of momentum is conserved:

$$P_x = MV + mv_x = 0 .$$

Furthermore, we have that:

$$M V_f + m v_f = 0 .$$

For the conservation of mechanical energy:

$$- \Delta E_p = \Delta E_k ,$$

Or:

$$mgR = m v_f^2 / 2 + M V_f^2 / 2 .$$

Combining the two relations, we have:

$$v_f^2 = 2gR \frac{M}{m + M} , \qquad V_f^2 = 2gR \frac{m^2}{M (m + M)} .$$

For the center of mass:

$$(v_{\text{cm}})_x = \frac{mv_x + MV}{m + M} = 0 , \qquad (v_{\text{cm}})_y = \frac{mv_y}{m + M} .$$

For the conservation of mechanical energy we have:

$$mgR \sin\theta = \frac{1}{2}MV^2 + \frac{1}{2}mv_x^2 + \frac{1}{2}mv_y^2 .$$

Or:

$$mgR \sin\theta = \frac{1}{2}\frac{m}{M}(m + M)v_x^2 + \frac{1}{2}mv_y^2 .$$

In addition it is valid:

$$v_x = v_y \tan\theta \frac{M}{m+M} .$$

For which:

$$gR \sin\theta = \frac{1}{2}v_y^2 \left(1 + \frac{M \tan^2\theta}{m+M}\right) ,$$

$$v_y^2 = 2gR \sin\theta \frac{m+M}{m+M+M \tan^2\theta} ,$$

And then:

$$(v_{cm})_y^2 = \left(\frac{m}{m+M}\right)^2 v_y^2 = 2gR \frac{m \sin\theta}{(m+M)(m+M+M \tan^2\theta)} .$$

Exercise 10

Three bodies are arranged as:

Collisions are elastic.
Find the value of mass2 that maximizes the velocity of body 3.

The laws of conservation of momentum and mechanical energy are translated as follows:

$$m_1 v_1 = m_1 v_1' + m_2 v_2' ,$$
$$m_1 v_1^2/2 = m_1 v_1'^2/2 + m_2 v_2'^2/2 .$$

Therefore:

$$\frac{m_1 + m_2}{m_1} v_2'^2 - 2 v_1 v_2' = 0 .$$

The non-trivial solution is:

$$v_2' = 2 v_1 \frac{m_1}{m_1 + m_2} ,$$

$$v_1' = \frac{m_1 - m_2}{m_1 + m_2} v_1 .$$

$$v_3' = 2 v_2' \frac{m_2}{m_2 + m_3} .$$

Therefore:

$$v_3' = 4\,v_1 \frac{m_1 m_2}{(m_1 + m_2)(m_2 + m_3)}\,.$$

By differentiating, we have:

$$\frac{dv_3'}{dm_2} = 4\,v_1\,m_1 \frac{m_1 m_3 - m_2^2}{[(m_1 + m_2)(m_2 + m_3)]^2}\,.$$

So the maximum is:

$$m_2 = \sqrt{m_1 m_3}\,.$$

Exercise 11

Consider a homogeneous sphere of radius R and mass M.
Calculate the moment of inertia of the sphere about an axis passing through its center.
Calculate the moment of inertia of the sphere about an axis tangent to it.

From the definition we have:

$$I_0 = I_z = \rho \int_V r^2\,dV\,,$$

By differentiating and integrating:

$$dI_z = 2\pi\rho\,dz \int_0^{r_z} r'^3 dr' = \frac{1}{2}\pi\rho r_z^4 dz\,.$$

Given that:

$$r_z^2 = R^2 - z^2\,,$$

Is found:

$$I_0 \;=\; \int_{-R}^{+R} \mathrm{d}I_z \;=\; \frac{\pi\rho}{2}\int_{-R}^{+R} r_z^4 \mathrm{d}z \;=\; \frac{\pi\rho}{2}\int_{-R}^{+R}(R^2-z^2)^2\mathrm{d}z \;=\; \frac{8\pi\rho R^5}{15}$$

Recalling the volume of a sphere, we can rewrite:

$$I_0 \;=\; 2MR^2/5\ .$$

For the tangent axis, it is obtained simply by applying Steiner's theorem:

$$I_1 \;=\; I_0 + MR^2 \;=\; 7MR^2/5\ .$$

Exercise 12

Given:

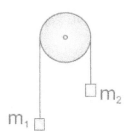

Determine the moment of inertia of the pulley knowing that the body 1 falls for a given height in a given time.
Determine the speed reached by the first body.

The laws of dynamics applied to the two bodies give:

$$m_1 a_1 \;=\; m_1 g + T_1\ , \qquad\qquad m_2 a_2 \;=\; m_2 g + T_2\ .$$

Also for the conservation of angular momentum:

$$I\alpha \ = \ \sum \tau \ = \ T_1 R - T_2 R \ .$$

But it is also true that:

$$a_1 \ = \ -a_2 \ , \qquad\qquad |a_1| \ = \ |a_2| \ = \ a \ .$$

$$a \ = \ \alpha R \ .$$

We therefore have the following system:

$$\begin{cases} m_1 a & = \ m_1 g - T_1 \\ -m_2 a & = \ m_2 g - T_2 \\ Ia/R & = \ (T_1 - T_2)R \end{cases}$$

From which:

$$T_1 \ = \ m_1 g - m_1 a \ , \qquad\qquad T_2 \ = \ m_2 g + m_2 a \ ,$$

Knowing also that, for the initial data:

$$a \ = \ 2y_0/t_0^2 \ ,$$

The moment of inertia of the pulley is obtained:

$$I \ = \ \left[(m_1 - m_2) \frac{g t_0^2}{2 y_0} \ - \ (m_1 + m_2) \right] R^2 \ .$$

The change in kinetic energy is:

$$\Delta E_k \ = \ m_1 v_1^2/2 \ + \ m_2 v_2^2/2 \ + \ I\omega^2/2 \ ,$$

While that of potential energy:

$$\Delta E_p \ = \ -(m_1 - m_2)\, g\, y_0 \ .$$

For the conservation of mechanical energy we have:

$$v = \sqrt{\frac{2gy_0 (m_1 - m_2)}{m_1 + m_2 + (I/R)^2}} \, .$$

Exercise 13

Given the system:

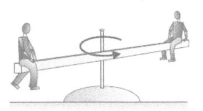

Where the carousel has a mass of 60 kg, the two people 30 kg, the length of the carousel is 2 meters and the carousel makes 1 turn every 2 seconds. Calculate the moment of inertia and the magnitude of the angular momentum of the system about the axis of rotation.
If people approach up to a distance of 80 cm, how has the system changed?
What is the change in mechanical energy of the system?

The moment of inertia of the member is:

$$I_a = \frac{1}{12} m (2d)^2 = \frac{1}{3} m d^2 = 20 \, \text{kg m}^2 \, .$$

That of the total system:

$$I_0 = I_a + 2Md^2 = 20 + 60 = 80 \, \text{kg m}^2 \, .$$

The angular speed of rotation is:

$$\omega_n = 2\pi\nu_n = 3.14\,\mathrm{rad\,s^{-1}}\ .$$

The modulus of angular momentum is:

$$L = I_0\,\omega_0 = 251.2\,\mathrm{kg\,m^2\,s^{-1}}\ .$$

If people get closer, the moment of inertia of the rod does not change, but the total one does:

$$I_1 = I_a + 2Md_1^2 = 20 + 38.4 = 58.4\,\mathrm{kg\,m^2}\ .$$

Angular momentum is conserved and so the angular velocity increases at:

$$\omega_1 = (I_0/I_1)\,\omega_0 = 4.3\,\mathrm{rad\,s^{-1}}\ .$$

The change in mechanical energy is:

$$\Delta E_k = I_1\omega_1^2/2 - I_0\omega_0^2/2 = 539.9 - 394.4 = 145\,\mathrm{J}\ .$$

Exercise 14

Given:

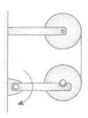

The two spools have the same mass M and the same radius R.
At a certain moment, the lower sprocket holder is removed and then it falls.
Find the acceleration of the falling reel and the tension in the rope holding the two reels together.
Is mechanical energy conserved?

The upper sprocket has a fixed axis and its motion is purely rotary. The equation of motion is:

$$I\,\alpha_1 \;=\; -T\,R\;,$$

The lower sprocket has a roto-translational motion. The equation of translational motion of the center of mass is:

$$M\,a_2 \;=\; M g\;-\;T\;.$$

While that for the rotational motion of the center of mass is:

$$I\,\alpha_2 \;=\; T\,R\;.$$

We obviously have:

$$\alpha_2 \;=\; -\alpha_1.$$

With respect to the inertial reference, the acceleration of the center of mass of the lower sprocket is:

$$a_2 \;=\; a' \;+\; a_t \;=\; 2\alpha R\;.$$

Therefore:

$$a_2 \;=\; g\left[\frac{I}{2MR^2}+1\right]^{-1}\;;\qquad T \;=\; Mg\left[\frac{1}{1+2MR^2/I}\right]^{-1}\;.$$

The elementary work done on the two spools is:

$$\mathrm{d}W_1 \;=\; T\,\mathrm{d}r_1 \;=\; \mathrm{d}E_{k,1}$$

$$\mathrm{d}W_2 \;=\; -T\,\mathrm{d}r_2 \;+\; M g\,\mathrm{d}r_{\mathrm{cm}} \;=\; \mathrm{d}E_{k,2}\;.$$

The change in kinetic energy is:

$$\Delta E_k \;=\; \mathrm{d}W_1 \;+\; \mathrm{d}W_2 \;=\; M g\,\mathrm{d}r_{\mathrm{cm}}$$

Depending only on the force of gravity, we are in a conservative field and therefore the mechanical energy is conserved.

Exercise 15

Given:

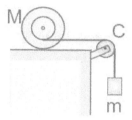

Where m=2 Kg, M=5 Kg, R= 0.2 m, r=0.1 m.
If the disk has pure rolling motion, with what acceleration does the body of mass m fall?
The tension T of the rope and the friction force are determined.

From the law of dynamics we have:

$$ma \; = \; mg - T \, .$$

The equation of motion of the disk in the case of pure rolling is:

$$I_Q \, \alpha \; = \; T(R - r) \, .$$

And then:

$$a \; = \; \alpha \, (R - r) \, .$$

From Steiner's theorem, the moment of inertia of the disk is:

$$I_Q \; = \; I_0 + MR^2 \; = \; 3MR^2/2 \, .$$

So the acceleration is:

$$a = g\left[\frac{3MR^2}{2m(R-r)^2} + 1\right]^{-1} = \frac{g}{16} = 0.613\,\mathrm{m\,s^{-2}}.$$

The tension in the rope is:

$$T = mg\left[1 + \frac{2m(R-r)^2}{3MR^2}\right]^{-1} = \frac{15}{16}mg = 18.4\,\mathrm{N}.$$

To calculate the frictional force, consider the equation of motion of the disk's center of mass:

$$Ma_{\mathrm{cm}} = T - F_a,$$

Where is it:

$$a_{\mathrm{cm}} = \frac{R}{R-r}a.$$

Therefore:

$$F_a = T - Ma_{\mathrm{cm}} = mg\frac{MR(R+2r)}{3MR^2 + 2m(R-r)^2} = 12.25\,\mathrm{N}.$$

Exercise 16

Given two rods arranged as in the figure:

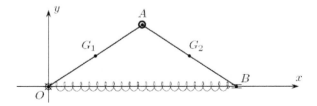

They are hinged at A. Rod OA is hinged at O with a smooth hinge while point B can slide on the axis with a rough constraint and a given coefficient of static friction.

An elastic force of origin O applied at point B and of constant k acts on the rods.

Determine the equilibrium positions of the system and the support reactions in O and B.

The system has one degree of freedom.

We assume as Lagrangian parameter the theta angle indicated in the figure of the acting forces:

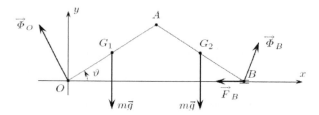

The coordinates of the points are given by:

$$- \ O: (0,0);$$
$$- \ G_1: (l/2 \cos \vartheta, l/2 \sin \vartheta);$$
$$- \ G_2: (3l/2 \cos \vartheta, l/2 \sin \vartheta);$$
$$- \ B: (2l \cos \vartheta, 0).$$

And the forces acting from:

$$- \ \vec{\Phi}_O = \Phi_{Ox}\vec{i} + \Phi_{Oy}\vec{j};$$
$$- \ \vec{\Phi}_B = \Phi_{Bx}\vec{i} + \Phi_{By}\vec{j};$$
$$- \ m\vec{g} = -mg\vec{j};$$
$$- \ \vec{F}_B = -\vec{F}_O = -2kl \cos \vartheta \vec{i}.$$

The resultant equation gives:

$$\Phi_{Ox} + \Phi_{Bx} - 2kl \cos \vartheta = 0,$$
$$\Phi_{Oy} + \Phi_{By} - 2mg = 0.$$

While the one about the moments:

$$\overrightarrow{OG_1} \wedge m\vec{g} + \overrightarrow{OG_2} \wedge m\vec{g} + \overrightarrow{OB} \wedge \vec{\Phi}_B = \vec{0},$$

From which:

$$\begin{vmatrix} \vec{\imath} & \vec{\jmath} & \vec{k} \\ 1/2\cos\vartheta & 1/2\sin\vartheta & 0 \\ 0 & -mg & 0 \end{vmatrix} + \begin{vmatrix} \vec{\imath} & \vec{\jmath} & \vec{k} \\ 3l/2\cos\vartheta & 1/2\sin\vartheta & 0 \\ 0 & -mg & 0 \end{vmatrix} + \begin{vmatrix} \vec{\imath} & \vec{\jmath} & \vec{k} \\ 2l\cos\vartheta & 0 & 0 \\ \Phi_{Bx} & \Phi_{By} & 0 \end{vmatrix} = \vec{0}.$$

And then:

$$- mg\cos\vartheta + \Phi_{By}\cos\vartheta = 0.$$

The rough constraint in B satisfies the Coulomb condition:

$$\left| \frac{\Phi_{Bx}}{\Phi_{By}} \right| \leq f_s.$$

And then:

$$-\frac{mg}{2}\cos\vartheta + \Phi_{By}\cos\vartheta + \Phi_{Bx}\sin\vartheta - 2kl\cos\vartheta\sin\vartheta = 0$$

The equilibrium positions are infinite as they are those for which the theta angle satisfies Coulomb's condition.

Exercise 17

A circular guide with center O and radius r is considered a rectilinear guide with equation x=2r in a three-dimensional plane.
A point P of mass m is bound, without friction, to the circular guide, while a point Q of mass M to the straight one.
An elastic force of constant h acts between the two points.
Find the equilibrium positions of the system.

116

The scheme is given by:

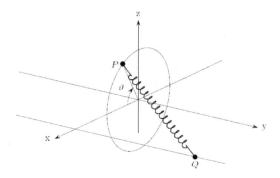

The scheme has two degrees of freedom and the Lagrangian parameters are the theta angle and the y coordinate.
The coordinates of the points and the acting forces are:

- $P(r \cos \vartheta, 0, r \sin \vartheta)$, $m\vec{g} = -mg\vec{k}$, $\vec{F}_P = h\vec{PQ} = (2hr - hr\cos\vartheta)\vec{i} + hy\vec{j} - hr\sin\vartheta\vec{k}$.

- $Q(2r, y, 0)$, $M\vec{g} = -Mg\vec{k}$, $\vec{F}_Q = -\vec{F}_P = (hr\cos\vartheta - 2hr)\vec{i} - hy\vec{j} + hr\sin\vartheta\vec{k}$.

While the virtual displacements of the two points of application of the forces are:

$$- \quad \delta P = -r\sin\vartheta\,\delta\vartheta\vec{i} + r\cos\vartheta\,\delta\vartheta\vec{k};$$
$$- \quad \delta Q = \delta y\vec{j}.$$

Equilibrium is obtained by setting the Lagrangian components of these stresses to zero.

Exercise 18

In a three-dimensional system, consider a line s in the xy plane free to rotate about O.
A homogeneous square plate with side l and mass m is constrained in the Osz plane with the center of gravity G on the straight line s, being able to rotate around G.

Determine the kinetic energy of the plate.

Let's take a triad with origin in G and two versors parallel to the sides of the lamina, while the third is orthogonal to the lamina.

The system has 3 degrees of freedom and as Lagrangian parameters we take the abscissa of the point G on the straight line s, the angle in the Oxy plane between the x axis and the straight line s, the angle in the Orz plane between the straight line s and the versor i .

The kinetic energy is given by:

$$K = K = \frac{1}{2} M \bar{v}_G{}^2 + \frac{1}{2} \left(Ap^2 + Bq^2 + Cr^2 \right) ,$$

We have:

$$\begin{cases} x_G & = & s \cos \varphi \\ y_G & = & s \sin \varphi \\ z_G & = & 0 \end{cases} ,$$

$$\begin{cases} \dot{x}_G & = & \dot{s} \cos \varphi - s \dot{\varphi} \sin \varphi \\ \dot{y}_G & = & \dot{s} \sin \varphi + s \dot{\varphi} \cos \varphi \\ \dot{z}_G & = & 0 \end{cases} ,$$

From which:

$$\bar{v}_G{}^2 = \dot{s}^2 + s^2 \dot{\varphi}^2 .$$

The angular velocity is:

$$\vec{\omega} = \dot{\vartheta} \vec{k} + \dot{\varphi} \vec{e}_3 ,$$

Being:

$$\vec{e}_3 = \sin \vartheta \vec{i} + \cos \vartheta \vec{j}.$$

We have:

$$\vec{\omega} = \dot\varphi \sin\vartheta \vec{i} + \dot\varphi \cos\vartheta \vec{j} + \dot\vartheta \vec{k} = p\vec{i} + q\vec{j} + r\vec{k}.$$

The moments of inertia are:

$$A = B = \frac{ml^2}{12} \quad , \quad C = A + B = \frac{ml^2}{6}.$$

So the kinetic energy is:

$$K = \frac{1}{2}m\left(\dot{s}^2 + s^2\dot\varphi^2\right) + \frac{1}{2}\left(\frac{ml^2}{12}\dot\varphi^2 \sin^2\vartheta + \frac{ml^2}{12}\dot\varphi^2 \cos^2\vartheta + \frac{ml^2}{6}\dot\vartheta^2\right) =$$
$$= \frac{1}{2}m\left(\dot{s}^2 + s^2\dot\varphi^2 + \frac{l^2}{12}\dot\varphi^2 + \frac{l^2}{6}\dot\vartheta^2\right).$$

Exercise 19

Given a square ABCD with side mass M=4m placed in a vertical plane Oxy, it is free to rotate around its vertex A by means of a smooth hinge.
It is subjected to an elastic force of constant h acting in C and with center in the projection H of C on the y axis.
Furthermore, there is a force of constant magnitude parallel and equivalent to AB acting on point B.
Determine the equilibrium positions, the reaction in A at equilibrium, the stability of the positions found, the kinetic energy of the system and the cardinal equations of the dynamics of the system.

The system has one degree of freedom and we take the angle in C as the Lagrangian parameter.
The coordinates are:

$$C = \left(l\sqrt{2}\cos\varphi, l\sqrt{2}\sin\varphi\right)$$
$$G = \left(\tfrac{l}{2}\sqrt{2}\cos\varphi, \tfrac{l}{2}\sqrt{2}\sin\varphi\right)$$
$$B = \left(l\cos\left(\varphi - \tfrac{\pi}{4}\right), l\sin\left(\varphi - \tfrac{\pi}{4}\right)\right)$$

The virtual displacements of these points where the forces are applied are:

119

$$\delta C = -l\sqrt{2}\sin\varphi\delta\varphi\vec{\imath} + l\sqrt{2}\cos\varphi\delta\varphi\vec{\jmath}$$
$$\delta G = -\tfrac{l}{2}\sqrt{2}\sin\varphi\delta\varphi\vec{\imath} + \tfrac{l}{2}\sqrt{2}\cos\varphi\delta\varphi\vec{\jmath}$$
$$\delta B = -l\sin\left(\varphi - \tfrac{\pi}{4}\right)\delta\varphi\vec{\imath} + l\cos\left(\varphi - \tfrac{\pi}{4}\right)\delta\varphi\vec{\jmath}$$

Where the acting forces are:

$$M\vec{g} = -Mg\vec{\jmath};$$
$$\overrightarrow{F}_{el} = h\overrightarrow{CH} = -hl\sqrt{2}\sin\varphi\vec{\imath};$$
$$\overrightarrow{F} = F\cos\left(\varphi - \tfrac{\pi}{4}\right)\vec{\imath} + F\sin\left(\varphi - \tfrac{\pi}{4}\right)\vec{\jmath}.$$

The stress work is given by:

$$\delta L^a = \left(-mg\frac{l}{2}\sqrt{2}\cos\varphi + 2kl^2\sin\varphi\cos\varphi\right)\delta\varphi = Q_\varphi\delta\varphi.$$

The equilibrium positions are found for:

$$-mg\frac{l}{2}\sqrt{2}\cos\varphi + 2kl^2\sin\varphi\cos\varphi = 0.$$

That is for:

$$\varphi_1 = \tfrac{\pi}{2},$$
$$\varphi_2 = -\tfrac{\pi}{2},$$
$$\varphi_3 = \arcsin\left(\frac{mg\sqrt{2}}{4hl}\right),$$
$$\varphi_4 = \pi - \arcsin\left(\frac{mg\sqrt{2}}{4hl}\right),$$

The potential energy is given by:

$$U = mgy_G + \frac{1}{2}h\|\overrightarrow{CH}\|^2 = mg\frac{l}{2}\sqrt{2}\sin\varphi + hl^2\cos^2\varphi.$$

$$Q_\varphi = -\frac{\partial U}{\partial x}.$$

Solutions are stable where the second derivative of U is positive. The kinetic energy is given by:

$$K = \frac{1}{2}\vec{\omega} \cdot \mathbb{I}_A(\vec{\omega}) = \frac{1}{2}\dot{\varphi}\vec{k} \cdot \mathbb{I}_A(\dot{\varphi}\vec{k}).$$

And then:

$$K = \frac{1}{2}(\ 0 \quad 0 \quad \dot{\varphi}\ ) \begin{pmatrix} I_{11} & I_{12} & I_{13} \\ I_{21} & I_{22} & I_{23} \\ I_{31} & I_{32} & I_{33} \end{pmatrix} \begin{pmatrix} 0 \\ 0 \\ \dot{\varphi} \end{pmatrix}$$

$$= \frac{1}{2}I_{33}\dot{\varphi}^2 = \frac{1}{2}\left(\frac{5}{6}Ml^2\right)\dot{\varphi}^2 = \left(\frac{5}{12}Ml^2\right)\dot{\varphi}^2.$$

The first cardinal equation of dynamics is:

$$\overrightarrow{R}^{\,\text{est}} = M\vec{a}_G,$$

Where is it:

$$\vec{v}_G = \frac{l}{2}\sqrt{2}\,(-\sin\varphi\,\dot{\varphi}\vec{\imath} + \cos\varphi\,\dot{\varphi}\vec{\jmath})$$

$$\vec{a}_G = \frac{l}{2}\sqrt{2}\,[(-\cos\varphi\,\dot{\varphi}^2 - \sin\varphi\,\ddot{\varphi})\,\vec{\imath} + (-\sin\varphi\,\dot{\varphi}^2 + \cos\varphi\,\ddot{\varphi})\,\vec{\jmath}]$$

The second cardinal equation of dynamics is:

$$\overrightarrow{M}_A^{\text{est}} = \dot{\overrightarrow{L}}_A.$$

$$\overrightarrow{L}_A = I_a\vec{\omega} = I\dot{\varphi}\vec{k}.$$

Or:

$$\overrightarrow{M}_A^{\text{est}} = I_a\ddot{\varphi}\vec{k},$$

5

CLASSICAL MECHANICS: THEORY OF GRAVITATION

Gravitation

Let us consider the theory of gravitation as derived from Newton's studies of classical mechanics.

The contemporary implications of the theory of gravitation are beyond the scope of this textbook.

The publication of Newton's theory coincided not only with the beginning of a scientific study of mechanics in all its forms, but also a general interpretation was given to the theory of gravitation and various astronomical problems, until then explained only by empirical observations or not totally scientific principles.

Newton established that two masses attract each other according to a force proportional to them and decreasing with the square of the distance.

$$\vec{F}_G = G \frac{m_1 m_2}{\vec{r}^2}$$

Where G is the universal gravitational constant whose value was calculated only in 1798 by Cavendish.

The fact that this value is infinitely small means that the gravitational force is not perceptible for bodies having a low mass, such as almost all physical bodies of daily experience.

Conversely, since this force is always attractive, as the masses are always positive, for highly massive bodies, such as planets and stars for example, the gravitational force becomes predominant.

This led Newton to state that the gravitational force is responsible both for the acceleration of gravity on planet Earth (the attraction that pushes the famous apple to fall to the ground) and for the mutual attraction between the planets.

To affirm all this, Newton had to assume two fundamental prerequisites:

1) The mass that appears in the second law of dynamics is called inertial mass, the one that appears in the law of universal gravitation is called gravitational. In Newton's theory it remains unclear why these two values always coincide, for any body. On a practical level, Newton took this assumption as valid and did not pose further problems.

2) The gravitational force is a force at a distance but which has an immediate effect on bodies, even in the astronomical field. This means that this force does not respect the Galilean principle of relativity. Newton did not ask other questions, syndicating that space and time were absolute.

The gravitational force is responsible for a gravitational field which is conservative and irrotational.
The potential of this force is given by:

$$V = -G\frac{m}{r}$$

Applying the well-known Gauss's theorem (unknown, however, in Newton's time!) we obtain:

$$\nabla^2 V = 4\pi G\rho$$

The acceleration due to gravity is defined as the value of the acceleration created in the earth's gravitational field when a mass is replaced by the mass of the earth and, at a distance, the earth's radius.
Precisely because of the values involved, the height at which an experiment is carried out for the purpose of determining the acceleration due to gravity is completely negligible.
On the other hand, not having la Terraa perfect spherical shape, there are different values of the acceleration of gravity if it is calculated at the poles rather than at the equator.
However, this diversity has repercussions on negligible decimal figures, therefore the acceleration of the earth's gravity can be considered as a constant.

Considering the mass of any body and equating the force of the second law of dynamics with that obtained from the law of universal gravitation, we obtain that the acceleration due to gravity is given by:

$$g = G\frac{M_T}{R_T^2}$$

Assuming that a body is thrown upwards, one can calculate the so-called escape velocity, i.e. the speed that it must possess initially in order to overcome the gravitational force and thus escape the gravitational field of the attractive body.

Considering, as in the case of the acceleration of gravity, the mass and the radius of the Earth, the earth's escape velocity can be calculated:

$$v_f = \sqrt{2G\frac{M_T}{R_T^2}}$$

The value of which is 11.2 km per second and is essential for space missions.

astronomical consequences

At an astronomical level, taking the gravitational potential as a reference, it can be seen that the angular momentum and the total mechanical energy are conserved.

The radial equation then becomes:

$$\left(\frac{dr}{dt}\right)^2 = \frac{2(W-V)}{m} - \left(\frac{L}{mr}\right)^2$$

Kepler's laws naturally follow from this equation.

Newton's mechanics therefore foresees Kepler's astronomical theory, with the planets revolving around the Sun following elliptical orbits in which the Sun is one of the foci.

It goes without saying that Newton's mechanics confirms Copernicus' heliocentric theories and fits in the wake of what Galileo already affirmed.

In particular, given ε the eccentricity of the elliptical orbit, a the length of the major semi-axis and b that of the minor semi-axis, considering the total masses of the system, the following quantities can be defined:

$$a = \frac{l}{1-\varepsilon^2} = \frac{k}{2W}$$

$$b = \sqrt{al}$$

$$\varepsilon^2 = 1 - \frac{l}{a}$$

$$l = \frac{L^2}{G\mu^2 M_{TOT}}$$

So the equation of the orbits becomes:

$$r(\vartheta) = \frac{l}{1+\varepsilon\cos(\vartheta - \vartheta_0)}$$

Five physical cases can be presented that describe five different conics:

1) k<0, eccentricity equal to zero: the orbit is circular
2) k<0, eccentricity between zero and one: the orbit is elliptical
3) k<0, eccentricity equal to one: the conic is parabolic
4) k<0, eccentricity greater than one: the conic is hyperbolic curved towards the center of force
5) k>0, eccentricity greater than one: the conic is hyperbolic curved against the center of force

Only in the first two cases does one have the constitution of a "linked" system, ie with one body that rotates around the other.
The last three cases, on the other hand, represent very well the trajectories of non-periodic comets, i.e. comets that do not follow a closed path on themselves, returning after a certain period to the same point.
Basically, the equation derived from Newtonian mechanics explains Kepler's first law and expands it in cases not yet studied.
Given C the position of the focus, the orbital surface covered between two instants of time is given by Kepler's second law, easily obtainable from the above theory with kinematic considerations:

$$A(t_1, t_2) = \frac{L_C}{2m}(t_2 - t_1)$$

Again from kinematic considerations, called T the period, we also derive Kepler's third law:

$$\frac{T^2}{a^3} = \frac{4\pi^2}{GM_{TOT}}$$

High school exercises

Exercise 1

Determine the value of the gravitational field generated by the Moon on its surface under the hypothesis that the mass of the Moon is homogeneously distributed and with spherical symmetry.

The radius and mass of the Moon are:

$$\frac{1,740 \cdot 10^6 \text{ m}}{7,35 \cdot 10^{22} \text{ Kg}}$$

The value of the gravitational field on the surface is:

$$g_L = G\frac{M_L}{R_L^2} = 6,67 \cdot 10^{-11} \frac{N \cdot m^2}{Kg^2} \cdot \frac{7,35 \cdot 10^{22} \text{ Kg}}{\left(1,740 \cdot 10^6\right)^2 \text{ m}^2} \approx 1,62\frac{m}{s^2}.$$

Exercise 2

Determine the ratio of the gravitational field of the earth to that of the moon.
Also determine the density of the Earth and the Moon.
Assume a spherical shape for both bodies.

The gravitational field of the earth at the surface is:

$$g_T = G\frac{M_T}{R_T^2} = 6,67\cdot10^{-11}\frac{N\cdot m^2}{Kg^2}\cdot\frac{5,97\cdot10^{24}\ Kg}{\left(6,378\cdot10^6\right)^2\ m^2} \simeq 9,79\frac{m}{s^2}$$

So the relationship with the lunar one is:

$$r = \frac{g_L}{g_T} = \frac{1,62}{9,79} \simeq \frac{1}{6}$$

Their respective densities are given by:

$$\rho_L = \frac{M_L}{\frac{4}{3}\pi R_L^3} = \frac{3}{4}\cdot\frac{7,35\cdot10^{22}\ Kg}{\pi\left(1,740\cdot10^6\right)^3} \simeq 3,33\cdot10^3\frac{Kg}{m^3}$$

$$\rho_T = \frac{M_T}{\frac{4}{3}\pi R_T^3} = \frac{3}{4}\cdot\frac{5,97\cdot10^{24}\ Kg}{\pi\left(6,378\cdot10^6\right)^3\ m^3} \simeq 5,496\cdot10^3\frac{Kg}{m^3}$$

Exercise 3

Determine the value of the gravitational field at half the Earth's and the Moon's radius.
Also determine the ratio between the two contributions.

The trend of the gravitational field inside the two bodies is:

$$g(P) = \frac{4}{3}\pi G\rho r,\ \text{con}\ 0\leq r\leq R,$$

The respective gravitational fields will therefore be:

$$g_T(P) = \frac{1}{2}\cdot\left(\frac{4}{3}\pi G\rho_T\cdot R_T\right) = 0,5\cdot9,79\frac{m}{s^2} \simeq 4,90\frac{m}{s^2}.$$

$$g_L(P) = \frac{1}{2}\cdot\left(\frac{4}{3}\pi G\rho_L\cdot R_L\right) = 0,5\cdot1,62\frac{m}{s^2} = 0,81\frac{m}{s^2}$$

Exactly half of those on the surface.
Their relationship therefore does not change.

Exercise 4

Calculate the escape velocity for the Earth and the Moon.

The escape velocity is given by:

$$v_f = \sqrt{2G\frac{M_T}{R_T^2}}$$

For the Earth it is 11.2 km per second.
For the Moon it is 1.82 km per second.
The ratio is always one-sixth.

University-level exercises

Exercise 1

Given two thin and homogeneous concentric spherical shells as in the figure:

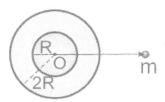

The smaller shell has radius R and mass M, the larger shell has radius 2R and mass 4M.
A point body of mass m is located at a distance r from the center O.
Determine the gravitational force for:

$$0 < r < R$$
$$R < r < 2R$$
$$r > 2R.$$

Also determine the dependence on the distance r of the gravitational potential energy.
If the mass m, initially at rest, is left free to move, with what speed will it reach the center of the shells?

Applying the law of universal gravitation, we have:

$$
\begin{aligned}
0 < r < R \qquad & F(r) = 0 \\
R < r < 2R \qquad & F(r) = \gamma mM/r^2 \\
r > 2R \qquad & F(r) = 5\gamma mM/r^2
\end{aligned}
$$

The gravitational potential energy is:

$$U(r) = -\int_{\infty}^{r} F(r) \cdot \mathrm{d}r .$$

Therefore:

$r > 2R$:

$$U(r) = 5\gamma mM \int_{\infty}^{r} \frac{\mathrm{d}r}{r^2} = -\frac{5\gamma mM}{r} ;$$

per $R < r < 2R$:

$$U(r) = \gamma mM \left[5\int_{\infty}^{2R} \frac{\mathrm{d}r}{r^2} + \int_{2R}^{r} \frac{\mathrm{d}r}{r^2} \right] = -\gamma mM \left(\frac{2}{R} + \frac{1}{r} \right) ;$$

per $0 < r < R$:

$$U(r) = -\gamma mM \left(\frac{2}{R} + \frac{1}{R} \right) = -\frac{3\gamma mM}{R} = \text{costante}.$$

For the conservation of mechanical energy on the mass m we will have:

$$\Delta E_k = -\Delta U .$$

And then:

$$\frac{1}{2} m v_0^2 = \frac{3\gamma m M}{R} \qquad \Rightarrow \qquad v_0 = \sqrt{\frac{6\gamma M}{R}} .$$

Exercise 2

A missile of mass m is launched from the earth's surface with an initial velocity greater than the escape velocity inclined at an angle theta to the horizontal.
The maximum distance reached by the missile is calculated as a function of the initial speed and theta.
Calculate the maximum distance for theta equal to 90° and 0°.

For the conservation of angular momentum we have:

$$m R v_0 \cos \theta = m r v .$$

And then:

$$\left(v_0^2 - \frac{2\gamma M}{R} \right) r^2 + 2\gamma r - R^2 v_0^2 \cos^2 \theta = 0 ,$$

From which:

$$r_{1,2} = \frac{-M \pm \sqrt{\gamma^2 M^2 + R^2 v_0^2 \cos^2 \theta \, (v_0^2 - 2\gamma M/R)}}{v_0^2 - 2\gamma M/R} ,$$

Of the two solutions, only the one with the minus sign before the radical should be taken.
If the angle is 90° we have:

$$r_{\max} = -\frac{2\gamma M}{v_0^2 - v_f^2} \, ,$$

Where the escape velocity appears in the denominator.
If it is 0° we have:

$$r_{1,2} = \frac{-M \pm \sqrt{\gamma^2 M^2 + R^2 v_0^2 \left(v_0^2 - 2\gamma M/R\right)}}{v_0^2 - 2\gamma M/R} \, .$$

6

FLUID THEORY AND FLUID DYNAMICS

Introduction

The study of fluid mechanics began much earlier than what was investigated with the scientific method.

We define pressure, the force per unit area.

In the International System, pressure is measured in Pascals (symbol Pa) and is equivalent to the force of one newton per square metre.

Other units of measurement used and accepted are the bar which is equal to 100,000 Pascals, the millibar which is equal to 100 Pascals, the atmosphere which is equal to 101,325 Pascals and the millimeter of mercury which is equal to 133.322 Pascals.

These units of measurement are somehow related to past experiments and everyday experience.

The atmosphere represents the pressure present at sea level and takes into account the column of air that weighs down on every surface of the earth.

As one moves to higher altitudes, this pressure decreases.

To this we owe various physical phenomena including the rarefaction of the air at high altitudes, the stratification of the atmosphere itself and the variation of the boiling temperature.

The millimeter of mercury instead derives from the experiment of Torricelli who, to measure the atmospheric pressure, used a basin containing mercury, proposing the equivalence between an atmosphere and 760 millimetrimercury.

We define density as the ratio of mass to volume.

A fluid is said to be compressible if its density can be changed as a result of the application of an external pressure. Conversely, it is said to be incompressible.

Stevino's law and Pascal's experiences

At the end of the sixteenth century, Stevino enunciated the law that bears his name, in which the pressure of a fluid with constant density depends only on the height of the fluid column:

$$p = \rho g h$$

This pressure accounts for the so-called Archimedean thrust, i.e. the force that a body immersed in a fluid receives and which allows it to float under appropriate conditions.

In 1653 Pascal enunciated a principle according to which the pressure is uniformly distributed on a surface level, through the famous experiment of compressed water in a circular container perforated in several points.

The water, subjected to the same force, comes out of each hole regardless of their arrangement, demonstrating how the pressure is distributed uniformly.

In 1644 Torricelli built the first instruments for measuring pressure and established the value of atmospheric pressure, which is added to the hydrostatic one if we want to consider the equations as a whole.

In fact, when we talk about pressures, we must always remember that there is always atmospheric pressure.

Precisely for this reason, the concepts of absolute pressure and relative pressure are used.

The absolute pressure is the one obtained by adding the atmospheric pressure to the hydrostatic one, the relative one is only given by the hydrostatic one.

Based on the density values, it can be seen that water has this characteristic: each amount 10 metriof water corresponds to an atmospheric pressure.

Therefore, a diver immersed at 40 metridepth will experience an absolute pressure of 5 atmospheres and a relative pressure of 4 atmospheres.

Bernoulli's principle

After the introduction of the scientific method and after the analyzes on mechanics conducted by Newton, it became evident how the laws of conservation of mass, mechanical energy and momentum could be extended to the theory of fluids.

In simplified cases, the so-called Bernoulli principle holds as an expression of the conservation laws.

For a stationary flow of an inviscid medium, this principle states that the following quantity is constant along a closed path:

$$\frac{1}{2}v^2 + gh + \int \frac{dp}{\rho}$$

For compressible fluids this quantity is constant along a streamline.

For isentropic processes and with irrotational speed, Bernoulli's principle is valid everywhere as it happens for incompressible fluids.

With this principle, a simple application of the conservation laws of Newtonian mechanics, we can for example calculate the speed of a fluid (for example water) in a pipeline after a drop in height (as occurs in hydroelectric plants), or calculate the pressure drop necessary to make the water rise to a certain height (as happened in multi-storey buildings, going to size the prevalence of the water booster pumps).

Rigorous definition of the theory of fluids

We give below the rigorous mathematical notation for the definition of pressure and for the fluid static and fluid dynamic equations.

Pressure is a scalar quantity.

The indefinite equation of hydrostatics is given by:

$$\nabla p = \vec{\rho} \cdot \vec{F}$$

The condition of hydrostatic equilibrium is obtained simply by saying that the sum of the mass forces and the forces acting on the surface is zero.

$$\int_V \rho F dV + \int_A p \bar{n} dA = 0$$

Where n denotes the direction perpendicular to surface A.

This condition of equilibrium derives directly from the concept of static equilibrium in mechanics and from Newton's third law of dynamics.

Pressure is simply defined as the derivative of force per unit area or as the derivative of energy per unit volume:

$$p = \frac{\partial F}{\partial A} = \frac{\partial E}{\partial V}$$

Mathematical tools

We introduce some useful theorems concerning nabla algebra and double or triple integrals.

Gauss's theorem relates the volumetric divergence to the surface integral:

$$\iiint (\nabla \cdot \vec{v})dV = \oiint (\vec{v} \cdot d\vec{n})d^2A$$

Stokes' theorem for a scalar considers the surface gradient and the line integral:

$$\iint (\vec{n} \times \nabla \phi)d^2A = \oint (\phi \cdot \vec{e}_t)ds$$

Stokes' theorem for a vector instead considers the surface curl and the line integral:

$$\iint \left[(\nabla \times \vec{v}) \cdot \vec{n}\right]d^2A = \oint (\vec{v} \cdot \vec{e}_t)ds$$

From simple mathematical extrapolations it can be stated that:

$$\oiint \left[(\nabla \times \vec{v}) \cdot \vec{n}\right]d^2A = 0$$

Ostrogradsky's theorems establish equalities for curls and volumetric divergences in case of vectors or scalars:

$$\iiint (\nabla \times \vec{v})d^3A = \oiint (\vec{n} \times \vec{v})d^2A$$
$$\iiint (\nabla \phi)d^3V = \oiint (\phi\vec{n})d^2A$$

In differential form, analogous theorems hold by recalling the algebra rules of the nabla operator.

Fluid dynamics

With these mathematical results, it is easy to pass from a static description to a dynamic one, giving rise to a particular sector of fluid mechanics, precisely called fluid dynamics.
The surface forces can be expressed in terms of the stress tensor T:

$$\vec{t} = \vec{n}T$$

If the viscous aspects can be ignored, this relation holds for the stress tensor:

$$divT = -\nabla p$$

Furthermore, if the fluid velocity can be separated into a translational part and a part responsible for rotations, deformations and expansions according to the following formula:

$$\vec{v}(d\vec{r}) = \vec{v}(\vec{r}) + d\vec{r} \cdot \nabla \vec{v}$$

the tensor given by the gradient of v, denoted as L, can be divided into a symmetric part, called D, and an antisymmetric part, called W as follows:

$$D_{ij} = \frac{1}{2}\left(\frac{\partial v_i}{\partial x_j} + \frac{\partial v_j}{\partial x_i} \right)$$

$$W_{ij} = \frac{1}{2}\left(\frac{\partial v_i}{\partial x_j} - \frac{\partial v_j}{\partial x_i} \right)$$

$$L = D + W = \nabla \vec{v}$$

Rotation defined as

$$\vec{\omega} = \nabla \times \vec{v}$$

137

we have the following relation for the antisymmetric part:

$$\bar{dr} \cdot W = \frac{1}{2} \omega \times \bar{dr}$$

The stress tensor can be expressed as the sum of the perpendicular stresses and the surface stresses (in this relation I is the unit tensor):

$$T = pI + T'$$

For Newtonian fluids the following relations hold between the surface stress tensor and D, the symmetrical part of the tensor given by the gradient of v:

$$T' = 2\eta D$$

$$\tau_{ij} = \eta \frac{\partial v_i}{\partial x_j}$$

where η is the dynamic viscosity of the fluid.
For a compressible fluid:

$$T' = (\eta' \nabla \cdot \bar{v})I + 2\eta D \Rightarrow 3\eta' + 2\eta = 0$$

If the dynamic viscosity is constant, the relations become:

$$div(2D) = \nabla^2 \bar{v} + \nabla(\nabla \cdot \bar{v})$$

Navier-Stokes equations

With the introduction of the laws for fluid mechanics, it becomes natural to extend and reformulate the principles of conservation of mass, energy and momentum, bringing everything back to an integral or differential notation.
For ease of writing, we only explain the second notation, remembering that, to go back to the first, it is enough to apply the mathematical results presented at the beginning of this chapter.

The conservation of mass is also called continuity equation and compared to the Newtonian vision it takes into account both a dynamic term due to the speed of the fluid and a static term given by the variation of density:

$$\nabla \cdot (\rho \vec{v}) + \frac{\partial \rho}{\partial t} = 0$$

In the static case, the conservation of mass implies that the density is constant over time (as already foreseen in mechanics).
On the other hand, if the density is constant over time, this equation generalizes a well-known principle according to which the flow rate of an incompressible fluid is constant.

Both the terms due to mechanics and those due to the stress tensor come into play for the conservation of momentum, including the part relating to pressure:

$$\rho \frac{d\vec{v}}{dt} + (\rho \vec{v} \cdot \nabla)\vec{v} = \vec{F} - \nabla p + divT'$$

For the conservation of energy, it is necessary to introduce thermodynamic concepts formalized during the nineteenth century, such as for example the entropy S and the heat flux q.
We note that the mere conservation of total mechanical energy is no longer sufficient to describe the complete physical system of mechanical, fluid and thermal parts.
In general terms, the conservation of energy is expressed as follows:

$$\frac{\rho T}{m} \frac{dS}{dt} = -\nabla \cdot \vec{q} + T' : D$$

Where la Tthe first member is the temperature.
For isentropic systems and without heat fluxes, i.e. for the mechanical systems studied so far, the conservation of energy is given by:

$$T' : D = 0$$

In this case, a more general form of energy conservation is expressed by considering also the electromagnetic contributions in the stress tensor:

$$\nabla_{v} T^{\mu\nu} = 0$$

By applying the conservation laws just described to an incompressible, viscous and heat conducting medium, the Navier-Stokes equations are obtained, which fully describe the fluid dynamics evolution of a medium and from which it is possible to obtain temperature, pressure, speed, capacity temperature and density of the fluid.
Defined C the thermal capacity and k the thermal conductivity, these equations are:

$$\nabla \cdot \bar{v} = 0$$

$$\rho \frac{\partial \bar{v}}{\partial t} + \rho(\bar{v} \cdot \nabla)\bar{v} = \rho \bar{g} - \nabla p + \eta \nabla^2 \bar{v}$$

$$\rho C \frac{\partial T}{\partial t} + \rho C(\bar{v} \cdot \nabla)T = \kappa \nabla^2 T + 2\eta D : D$$

These equations are nothing more than the conservation laws just explained.

The Navier-Stokes equations can be expressed with dimensionless quantities, once the numbers have been defined which are also useful for describing the practical applications of fluid dynamics.
Call c the speed of sound, L a generic length and define the following quantities:

kinematic viscosity

$$\nu = \frac{\eta}{\rho}$$

heat transfer coefficient

$$\alpha \rightarrow \kappa \hat{\partial} T = \alpha \Delta T$$

thermal diffusion coefficient

$$a = \frac{\kappa}{\rho c}$$

the following dimensionless numbers can be associated.

The Strouhal number establishes the ratio between the non-stationary and stationary inertial forces:

$$Sr = \frac{\omega L}{v}$$

The Froude number is the ratio of the stationary inertial forces to the force of gravity:

$$Fr = \frac{v^2}{gL}$$

The Mach number expresses the speed of the vehicle relative to the speed of sound.
If it exceeds 0.8 it causes shock waves that propagate in the medium.
If it is greater than 1, we speak of ultrasonic speeds.
This number is particularly important in aerodynamics.

$$Ma = \frac{v}{c}$$

The Fourier number takes into account thermal relationships between the heat conductivity and the energy change of the medium:

$$Fo = \frac{a}{\omega L^2}$$

The Eckert number is the ratio of viscous heat dissipation to convective heat transport:

$$Ec = \frac{v^2}{c\Delta T}$$

The Prandtl and Nusselt numbers take into account the specificities of the media materials:

$$Pr = \frac{\nu}{a}$$

$$Nu = \frac{L\alpha}{\kappa}$$

The Peclet number is the ratio of the convective heat transfer to the thermal conductivity:

$$Pe = \frac{vL}{a}$$

Finally, the Reynolds number expresses the ratio between stationary inertial forces and viscous forces:

$$\text{Re} = \frac{vL}{v}$$

With such numbers, the dimensionless Navier-Stokes equation is as follows:

$$Sr \frac{\partial \bar{v}'}{\partial t'} + (\bar{v}' \cdot \nabla')\bar{v}' = -\nabla' p + \frac{\bar{g}}{Fr} + \frac{\nabla'^2 \bar{v}'}{\text{Re}}$$

In which also the quantities with the prime are parameterized as follows:

$$x' = \frac{x}{L}$$

$$\bar{v}' = \frac{\bar{v}}{V}$$

$$\nabla' = L\nabla$$

$$\nabla'^2 = L^2\nabla^2$$

$$t' = t\omega$$

Applications of the Navier-Stokes equations

For tubular flows, if L is considered as the diameter of the pipe, the flow is laminar if the Reynolds number is less than 2300, while it is turbulent if it is greater than this value.

For an incompressible laminar flow in a circular pipe, the velocity profile is given by:

$$\Phi_V = -\frac{\pi}{8\eta}\frac{dp}{dx}R^4$$

Still in the case of tubular flows, for the transport of gas at low pressures, this profile is:

$$\Phi_V = \frac{4R^3\alpha\sqrt{\pi}}{3}\frac{dp}{dx}$$

Finally, for low values of the Reynolds number in tubular flows, the following relations hold:

$$\nabla p = \eta\nabla^2\vec{v}$$
$$\nabla\cdot\vec{v} = 0$$

For fluid dynamics, a theory of potentials can be defined as already done in mechanics.

Define the circulation along a closed line as:

$$\Gamma = \oint(\vec{v}\cdot\vec{e}_t)ds$$

For an inviscid medium, with a conservative force field and with the pressure that depends only on the density of the medium, we have the following Kelvin theorem:

$$\frac{d\Gamma}{dt} = 0$$

For an irrotational flow, one can express the velocity as the gradient of a potential and one can rewrite the Navier-Stokes equation for the potentials which, however, can be solved only in simplified cases.

Two examples are given by vortices and uniform media with high values of the Reynolds number.

In the first case the potential is given by:

$$\phi = \frac{\Gamma \vartheta}{2\pi}$$

In the second case, the forces respect these relations:

$$F_x = 0$$
$$F_y = -\rho \Gamma v$$

In all other cases it is necessary to define the boundary conditions for the flow and for the temperature in order to derive the solutions of these equations (which are partial derivatives and therefore solvable in analytical form only through the separation of variables. A different and very more powerful is given by the numerical resolution of the same).

For temperature boundary conditions, if the thickness of this boundary is much smaller than L, two cases can be obtained based on the value of the Prandtl number.

$$\Pr \leq 1 \Rightarrow \frac{\delta}{\delta_T} \approx \sqrt{\Pr}$$

$$\Pr \gg 1 \Rightarrow \frac{\delta}{\delta_T} \approx \sqrt[3]{\Pr}$$

For the flow boundary conditions, if the thickness of this boundary is much smaller than L, the following relationship holds:

$$\delta \approx \frac{L}{\sqrt{\mathrm{Re}}}$$

For the layer considered at the boundary there is the Von Karman equation of moments:

$$\frac{d}{dx}(\mathcal{G}v^2) + \delta^* v \frac{dv}{dx} = \frac{\tau_0}{\rho}$$

Having defined the following quantities:

$$\mathcal{G}v^2 = \int_0^\infty (v - v_x)v_x \, dy$$

$$\delta^* v = \int_0^\infty (v - v_x) \, dy$$

$$\tau_0 = -\eta \frac{\partial v_x}{\partial y}\Big|_{y=0}$$

For turbulent flows, the Navier-Stokes equations become:

$$\frac{\partial \langle \bar{v} \rangle}{\partial t} + (\langle \bar{v} \rangle \cdot \nabla)\langle \bar{v} \rangle = -\frac{\nabla \langle p \rangle}{\rho} + v\nabla^2 \langle \bar{v} \rangle + \frac{div(S_R)}{\rho}$$

Where the mean values are indicated in brackets (the only ones that count in turbulent flows) and S is the turbulent stress tensor for which an equality similar to that of Newtonian fluids holds:

$$S_R = -\rho \langle v_i v_j \rangle = 2\rho v_t \langle D \rangle$$

The boundary conditions for flow are different when considered close to the boundary or away from it.
In the first case it is

$$v_t = 0$$

In the second however

$$v_t \approx v\text{Re}$$

Through these equations it is possible to describe the mechanics of fluids in any existing physical situation, generating an almost unlimited series of practical applications.

The great theoretical and scientific innovations therefore led to the creation of new technological devices that significantly changed daily life, especially during the eighteenth and nineteenth centuries.

A new impetus to these studies was given in the contemporary era by aeronautics and space missions, as well as by applications to new types of fluid compared to those previously studied (water and the main gases).

Finally, we point out that the resolution of the Navier-Stokes equations in exact form is a very rare fact.

In the vast majority of cases, the equations are solved in numerical form only.

For this reason, numerical calculation and numerical analysis techniques find great application in fluid dynamics.

High school exercises

Exercise 1

In a horizontal pipe with a section of 10 square centimeters water flows at a speed of 8 meters per second with a pressure of 150,000 Pascals.
The section increases up to 16 square centimeters.
Calculate the velocity and pressure in this section.

Applying the law of conservation of flow (derived from the law of conservation of mass) we have:

$$S_1 V_1 = S_2 V_2$$

$$V_2 = \frac{S_1 V_1}{S_2} = \frac{10\,cm^2 \cdot 8\,\frac{m}{s}}{16\,cm^2} = 5\,\frac{m}{s}$$

Bernoulli's law says that:

$$\frac{1}{2}\rho_{H_2O} V_1^2 + \rho_{H_2O} g h_1 + P_1 = \frac{1}{2}\rho_{H_2O} V_2^2 + \rho_{H_2O} g h_2 + P_2$$

But the tube is horizontal, so the heights are the same, so:

$$\frac{1}{2}\rho_{H_2O}V_1^2 + P_1 = \frac{1}{2}\rho_{H_2O}V_2^2 + P_2$$

From simple counts and substituting what was found before:

$$\frac{1}{2}\rho_{H_2O}V_1^2 + P_1 = \frac{1}{2}\rho_{H_2O}V_1^2\frac{S_1^2}{S_2^2} + P_2$$

$$\frac{1}{2}\rho_{H_2O}V_1^2 - \frac{1}{2}\rho_{H_2O}V_1^2\frac{S_1^2}{S_2^2} + P_1 = P_2$$

$$P_2 = \frac{1}{2}\rho_{H_2O}V_1^2\left(1 - \frac{S_1^2}{S_2^2}\right) + P_1$$

Therefore:

$$P_2 = \frac{1}{2}1000\frac{kg}{m^3}64\frac{m^2}{s^2}\left(1 - \frac{100\,cm^4}{256\,cm^4}\right) + 150000\,Pa$$
$$P_2 = 169500\,Pa$$

Exercise 2

In a horizontal tube with a section of 10 square centimeters water flows at a speed of 3 meters per second.
The section decreases to 4 square centimeters.
How much is the range worth?
And the speed of the water in the section of 4 square centimeters?

The flow rate of the tube is:

$$Q = S_1 \cdot V_1 = 10\,cm^2 \cdot 3\frac{m}{s} = 0,001\,m^2 \cdot 3\frac{m}{s} = 0,003\frac{m^3}{s}$$

By the law of conservation of flow:

147

$$S_2 \cdot V_2 = S_1 \cdot V_1$$

$$V_2 = \frac{S_1 \cdot V_1}{S_2} = \frac{10\,cm^2 \cdot 3\,\frac{m}{s}}{4\,cm^2} = 7,5\,\frac{m}{s}$$

Exercise 3

A cylinder with a section of 10 square centimeters contains water up to a certain level.
Applying a hole with a 1 square millimeter section at 40 cm from the water level, how fast does the liquid come out?

From the law of conservation of flow we have:

$$S_1 V_1 = S_2 V_2$$

$$V_1 = \frac{S_2 V_2}{S_1}$$

And from Bernoulli's law:

$$\frac{1}{2}\rho_{H_2O} V_1^2 + \rho_{H_2O}\,gh_1 + P_1 = \frac{1}{2}\rho_{H_2O} V_2^2 + \rho_{H_2O}\,gh_2 + P_2$$

The pressures are the same and therefore not included in the calculation, as is the density of the water:

$$\frac{1}{2}V_1^2 + gh_1 = \frac{1}{2}V_2^2 + gh_2$$

Substituting what was previously found, we have:

$$\frac{1}{2}\frac{S_2^2 V_2^2}{S_1^2} - \frac{1}{2}V_2^2 = gh_2 - gh_1$$

And then:

$$V_2 = \sqrt{\frac{2g\Delta h}{1 - \frac{S_2^2}{S_1^2}}}$$

I our case:

$$V_2 = \sqrt{\frac{2 \cdot 9,8\frac{m}{s^2} \cdot 0,4\,m}{1 - \frac{1\,mm^2}{10\,cm^2}}} = \sqrt{\frac{7,84\frac{m^2}{s^2}}{1 - \frac{1}{1000}}} = 2,8\frac{m}{s}$$

Exercise 4

A U-shaped pipe contains water on the left and oil on the right.
The liquids are still.
Knowing that the height of the oil column is 20 cm, how many centimeters
is it higher than that of water?

The pressures of the two columns are:

$$P_{H_2O} = P_{atm} + \rho_{H_2O}g\Delta h_{H_2O}$$
$$P_{olio} = P_{atm} + \rho_{olio}g\Delta h_{olio}$$

And evidently they must be identical, as the liquids are still.
For which:

$$P_{atm} + \rho_{H_2O}g\Delta h_{H_2O} = P_{atm} + \rho_{olio}g\Delta h_{olio}$$

$$\rho_{H_2O}\Delta h_{H_2O} = \rho_{olio}\Delta h_{olio}$$

And then:

$$\Delta h_{H_2O} = \frac{\rho_{olio}\Delta h_{olio}}{\rho_{H_2O}} = 16\,cm$$

The oil column is higher than 4cm.

Exercise 5

The two sections of a hydraulic wine press are 50 and 5 square centimeters. Knowing that an object with a mass of 50 kg is supported in the largest section, what force must I exert on the second section to maintain balance?

Equilibrium occurs when the pressures are equal:

$$P_2 = P_1$$
$$\frac{F_2}{S_2} = \frac{F_1}{S_1}$$

And then:

$$F_2 = \frac{mg \cdot S_2}{S_1} = \frac{50\,kg \cdot 9,8\,\frac{m}{s^2} \cdot 5\,cm^2}{50\,cm^2} = 49\,N$$

Exercise 6

Mercury, water and oil are poured into a vertical cylinder.
The columns of each liquid are 5, 20 and 15 centimeters high, respectively.
Find the pressure at the bottom of the column.

The separation lines between the various liquids are found at:

$$h_a = h_0 - L_{olio} = -15\,cm$$
$$h_b = h_a - L_{H_2O} = -35\,cm$$
$$h_c = h_b - L_{Hg} = -40\,cm$$

Having fixed the zero at the top of the column.
The section in oil will have a pressure of:

$$\Delta P_{0 \to a} = -\rho g \Delta h_{0 \to a} = -\rho g \left(h_a - h_0 \right)$$

$$\Delta P_{0 \to a} = -800 \, \frac{kg}{m^3} \cdot 9,8 \, \frac{m}{s^2} \cdot (-15 \, cm - 0 \, cm) = 1176 \, Pa$$

The stretch in the water:

$$\Delta P_{a \to b} = -\rho g \Delta h_{a \to b} = -\rho g \left(h_b - h_a \right)$$

$$\Delta P_{a \to b} = -1000 \, \frac{kg}{m^3} \cdot 9,8 \, \frac{m}{s^2} \cdot (-35 \, cm + 15 \, cm) = 1960 \, Pa$$

And the mercury one:

$$\Delta P_{b \to c} = -\rho g \Delta h_{b \to c} = -\rho g \left(h_c - h_b \right)$$

$$\Delta P_{b \to c} = -13579 \, \frac{kg}{m^3} \cdot 9,8 \, \frac{m}{s^2} \cdot (-40 \, cm + 35 \, cm) = 6653,71 \, Pa$$

The total pressure will be the sum of these three pressures plus the atmospheric one:

$$P_c = P_0 + \Delta P_{0 \to a} + \Delta P_{a \to b} + \Delta P_{b \to c}$$

$$P_c = 100000 \, Pa + 1176 \, Pa + 1960 \, Pa + 6653,71 \, Pa = 109789,71 \, Pa$$

Exercise 7

A cylindrical container is filled with water up to a height of 30 centimeters from the bottom.
A hole of negligible size is drilled 5 cm from the bottom.
How fast does the water leave the hole?

From the Bernoulli equation:

$$\frac{1}{2}\rho V_f^2 + \rho g h_f + P_f = \frac{1}{2}\rho V_i^2 + \rho g h_i + P_i$$

But the pressures are identical (and are equal to atmospheric pressure) so they cancel out:

$$\frac{1}{2}\rho V_f^2 - \frac{1}{2}\rho V_i^2 = \rho g h_i - \rho g h_f$$

From the law of conservation of flow:

$$S_i V_i = S_f V_f$$

We have:

$$V_i = \frac{S_f}{S_i} V_f$$

And then:

$$V_f = \sqrt{\frac{\rho g (h_i - h_f)}{\frac{1}{2}\rho \left(1 - \frac{S_f^2}{S_i^2}\right)}}$$

Since the size of the hole is negligible, the relation becomes:

$$V_f = \sqrt{\frac{g (h_i - h_f)}{\frac{1}{2}}} = \sqrt{2g\Delta h} = 2,21 \frac{m}{s}$$

University-level exercises

Exercise 1

A homogeneous ball of volume V is suspended by a string from the lid of a container filled with air.
The container slides without friction on an inclined plane at an angle of theta to the horizontal.
Suppose the density of the ball is greater than that of the air.
Calculate the forces acting on the ball.
Determine the angle of inclination of the ball as the container descends the inclined plane.

The weight force acts on the ball:

$$P = \rho_1 V g;$$

Archimedean thrust:

$$S = -\rho_0 V g;$$

The tension of the thread T, the drag force and the Archimedean force:

$$F = -\rho_1 V a_0;$$
$$R = \rho_0 V a_0.$$

At equilibrium we have:

$$\rho_1 V a_0 = P + S + T + R.$$

With respect to the horizontal and vertical axes we have:

$$x' : \quad T_x - \rho_1 V a_0 \cos\theta + \rho_0 V a_0 \cos\theta = 0,$$
$$y' : \quad T_y - \rho_1 V g + \rho_0 V g + \rho_1 V a_0 \sin\theta - \rho_0 V a_0 \sin\theta = 0.$$

But:

153

$$a_0 = g \sin \theta$$

So:

$$
\begin{cases}
T_x = (\rho_1 - \rho_0)\, Vg \,\sin \theta \cos \theta \,, \\[2mm]
T_y = (\rho_1 - \rho_0)\, Vg \,(1 - \sin^2 \theta) \,.
\end{cases}
$$

So the angle will be:

$$\tan \alpha = \frac{T_y}{T_x} = \frac{1 - \sin^2 \theta}{\sin \theta \cos \theta} = \cot \theta$$

$$\alpha = \pi/2 - \theta \,.$$

Exercise 2

A ball of mass m is dropped with a given initial velocity in a liquid of given viscosity.
Determine the speed of the ball as a function of time.

The weight force and the viscous friction force act on the ball.
The law of motion is:

$$ma = mg - k\eta v \,.$$

From which:

$$m \frac{dv}{dt} = mg - k\eta v \,.$$

Place:

$$\gamma = k\eta/m,$$
$$w = g - \gamma v,$$

The solution of the equation is:

$$w = w_0\, e^{-\gamma(t-t_0)}.$$

Therefore:

$$v(t) = \frac{mg}{k\eta} - \left(\frac{mg}{k\eta} - v_0\right) \exp\left[-\frac{k\eta}{m}(t - t_0)\right].$$

Exercise 3

Considering a fluid in a static situation, subject only to the weight force, find the resultant force that the outside of a fluid region P exerts on the region itself.

It is about deriving Archimedes' principle.
Given the force field of the weight force:

$$b = -g e_3$$

The resultant force is the volume integral:

$$F = \int_P \rho b\, dV = -g \left(\int_P \rho\, dV\right) e_3 = -gM(P)e_3,$$

Exercise 4

Find the pressure range and the velocity range:

$$v(x, y, t) = u(x, y, t)e_1$$

Of a perfect incompressible fluid moving on the plane between the straight lines y=0 and y=h, in the absence of external forces and with the boundary conditions:

$$p(0, y, t) = p_1$$
$$p(L, y, t) = p_2$$

From the continuity equation:

$$\frac{\partial u}{\partial x} = 0$$

We have u=u(y,t).
From the second Euler equation:

$$\frac{\partial p}{\partial y} = 0$$

We have p=p(x,t).
While the first Euler equation:

$$\rho_0 \frac{\partial u}{\partial t} = -\frac{\partial p}{\partial x}$$

Returns:

$$\frac{\partial^2 p}{\partial x^2} = 0,$$

And then:

$$p(x, t) = A(t)x + B(t)$$

With the boundary conditions we have:

$$B(t) = p_1, \qquad A(t) = \frac{p_2 - p_1}{L}$$

From which by replacing:

$$p(x) = \frac{p_2 - p_1}{L} x + p_1 .$$

$$\frac{\partial u}{\partial t} = -\frac{p_2 - p_1}{\rho_0 L} ,$$

$$u(y, t) = -\frac{p_2 - p_1}{\rho_0 L} t + u_0(y)$$

Exercise 5

Find the velocity field profile for the motion:

$$\mathrm{grad}\, p(x, y, z) = C e_z .$$

In a tube with an elliptical section.

The Cartesian equation of an ellipse is:

$$\frac{x^2}{a^2} + \frac{y^2}{b^2} = 1 .$$

The speed range will be:

$$v(x, y) = w(x, y) e_z$$

And the equation of motion is:

$$\mu \Delta w = C,$$

On the ellipse we have the following boundary condition:

$$w(x, y) = 0$$

Place:

$$x' = x, \ y' = \tfrac{a}{b} y,$$

$$r = \sqrt{x'^2 + y'^2},$$

Is found:

$$\frac{\partial^2 w}{\partial x'^2} + \frac{a^2}{b^2} \frac{\partial^2 w}{\partial y'^2} = \frac{C}{\mu},$$

Whose solutions with cylindrical symmetry are:

$$w(r) = c_1(r^2 - a^2) = c_1(x'^2 + y'^2 - a^2)$$

From the boundary condition we have:

$$c_1 = \frac{C}{2\mu} \frac{b^2}{a^2 + b^2}.$$

And then:

$$w(x, y) = \frac{C}{2\mu} \frac{b^2}{a^2 + b^2} \left(x^2 + \frac{a^2}{b^2} y^2 - a^2 \right) = \frac{C}{2\mu} \frac{a^2 b^2}{a^2 + b^2} \left(\frac{x^2}{a^2} + \frac{y^2}{b^2} - 1 \right),$$

Exercise 6

Find the dispersion law for the phase velocities of the normal perturbation modes for the steady state velocity profile:

$$U(y) = \begin{cases} 1 & \text{se} \quad 1 < y < y_0 \\ y & \text{se} \quad -1 < y < 1 \\ -1 & \text{se} \quad -y_0 < y < -1 \end{cases}$$

$$y_0 = +\infty$$

Recalling that the Rayleigh equation for phase velocity is:

$$(U - c)\left(\frac{d^2}{dy^2} - k^2\right)\hat{\psi} - \frac{d^2 U}{dy^2}\hat{\psi} = 0,$$

The regions associated with solutions of the Rayleigh equation are:

$$\hat{\psi}^{(j)} = A_-^{(j)} e^{-ky} + A_+^{(j)} e^{ky} \quad j = 0, 1, 2.$$

In our case we have:

$$A_-^{(0)} = A_+^{(2)} = 0.$$

Imposing the conditions of continuity:

$$\left[\frac{\hat{\psi}^{(j)}}{a^{(j)} + b^{(j)}y - c}\right]_{y=y_j+1} = \left[\frac{\hat{\psi}^{(j+1)}}{a^{(j+1)} + b^{(j+1)}y - c}\right]_{y=y_j+1}$$

Is found:

$$A_-^{(1)}e^{-k} + A_+^{(1)}e^k - (1-c)(-kA_-^{(1)}e^{-k} + kA_+^{(1)}e^k) = -(1-c)(-kA_-^{(2)}e^{-k})$$
$$A_-^{(1)}e^k + A_+^{(1)}e^{-k} - (-1-c)(-kA_-^{(1)}e^k + kA_+^{(1)}e^{-k}) = -(-1-c)(kA_+^{(0)}e^{-k})$$
$$A_-^{(1)}e^{-k} + A_+^{(1)}e^k = A_-^{(2)}e^{-k}$$
$$A_-^{(1)}e^k + A_+^{(1)}e^{-k} = A_+^{(0)}e^{-k}$$

From which by replacing:

$$A_-^{(1)} + (1 - 2k(1 - c))A_+^{(1)}e^{2k} = 0$$
$$(1 - 2k(1 + c))A_-^{(1)}e^{2k} + A_+^{(1)} = 0$$

And so the non-trivial solutions are:

$$c^2 = (1 - 2k)^2 - e^{-4k}.$$

There will certainly be some unstable modes as the second side can be positive or negative according to k.

Exercise 7

Given a turbulent motion in a body with a diameter of 300 m, assuming that the speed of the vortices is 3 m/s and that the viscosity and density are:

$$\nu = 15 \cdot 10^{-6} \text{ m}^2/\text{s}$$
$$\rho = 1.25 \text{ kg/m}^3$$

Calculate the characteristic Reynolds number, give an estimate for the rate of energy dissipated per unit mass and the total power dissipated.

The Reynolds number is obtained from the definition:

$$R_0 = \frac{v_0 \ell_0}{\nu} = \frac{3 \cdot 300}{15 \cdot 10^{-6}} = 0.6 \cdot 10^8,$$

The energy dissipated per unit mass is:

$$\epsilon \approx \frac{v_0^3}{\ell_0} = \frac{27 \text{ m}^2}{300 \text{ s}^3} = 0.09 \text{ W/kg}.$$

And the power dissipated by the inertial motions is:

$$P = \epsilon\rho\ell_0^3 = 0.09 \cdot 1.25 \cdot 27000000 \text{ W} = 3.0375 \cdot 10^6 \text{ W} \,;$$

7

OPTICS

Introduction

The study of optical properties was already explored in antiquity, particularly in Greek times.

One of the fundamental assumptions coming from the ancient world, in particular from Euclid's studies, was that of the rectilinear propagation of light rays; this is the approximation of geometric optics that we will use in this section.

In modern times, the first scientist who ventured into the study of optics was Kepler who in 1610 tried to give a mathematical sense to the properties of lenses.

Subsequently Snell derived the laws of refraction, which still bear his name today.

In 1665 Grimaldì observed the phenomenon of diffraction, while Huygens in 1690 and Newton in 1704 proposed a wave nature for light, in opposition to the dominant corpuscular theory of the time.

Subsequent studies by Young on interference (1801), by Fresnel and by Foucault for measuring the speed of light brought some evidence about the wave nature without completely discarding the corpuscular one.

Let's say right away that the reconciliation of these two natures will find explanation only with contemporary physics of the twentieth century (in particular with quantum mechanics) and that the approximation of geometric optics will be overcome at the end of the nineteenth century with the theory of electromagnetism .

To understand the laws of optics, we need to introduce two quantities typically associated with wave phenomena.

We define wavelength as λ the spatial distance between two points of a wave in which there is the same value of a given quantity:

$$f(x) = f(x + \lambda)$$

In the same way we define the frequency f as the inverse of a time distance between two points of a wave in which there is the same value of a given quantity:

$$g(t) = g(t + \frac{1}{f})$$

For a wave, the following relationship holds, where v is the velocity:

$$\lambda f = v$$

Laws of refraction and reflection

Given a light ray which passes through two different mediums, we call the ray which propagates in the second medium refracted, while the ray which remains in the first medium is reflected.
Snell's law of refraction explains the deflection of light rays when passing from one material to another, by introducing a quantity called the refractive index:

$$n_i \sin \vartheta_i = n_t \sin \vartheta_t$$

Where the subscripts i and t refer to the incident light (before the separation surface between the different materials) and to the transmitted light (after the separation surface).
The following relation holds for refractive indices:

$$\frac{n_t}{n_i} = \frac{\lambda_i}{\lambda_t} = \frac{v_i}{v_t} = \frac{f_t}{f_i}$$

The importance of the refractive index is given by the fact that, by applying the mechanical concepts of minimum action and variational calculus, Hamilton's principle translates, in optics, into the famous Fermat's principle, according to which the trajectory of a ray of light in geometric optics is determined by the following result:

$$\delta \int_{1}^{2} dt = \delta \int_{1}^{2} n(s)ds = 0$$

When a light ray crosses two different materials, not only the refraction mechanism between the incident ray and the transmitted ray is generated, but there is also a reflection of the incident ray itself.
Snell's law of reflection states that the reflected angle is equal to the incident angle:

$$\vartheta_r = \vartheta_i$$

We can geometrically decompose the reflected and transmitted part in terms of component parallel and component perpendicular to the separating surface:

$$r_{\parallel} = \left(\frac{A_r}{A_i} \right)_{\parallel}$$

$$r_{\perp} = \left(\frac{A_r}{A_i} \right)_{\perp}$$

$$t_{\parallel} = \left(\frac{A_t}{A_i} \right)_{\parallel}$$

$$t_{\perp} = \left(\frac{A_t}{A_i} \right)$$

Where A is the amplitude of the wave related to the light beam.
Fresnel's equations express the relationships between these quantities and the angles of incidence and transmission:

165

$$r_{\parallel} = \frac{\tan(\vartheta_i - \vartheta_t)}{\tan(\vartheta_i + \vartheta_t)}$$

$$r_{\perp} = \frac{\sin(\vartheta_t - \vartheta_i)}{\sin(\vartheta_t + \vartheta_i)}$$

$$t_{\parallel} = \frac{2\sin\vartheta_t \cos\vartheta_i}{\sin(\vartheta_t + \vartheta_i)\cos(\vartheta_t - \vartheta_i)}$$

$$t_{\perp} = \frac{2\sin\vartheta_t \cos\vartheta_i}{\sin(\vartheta_t + \vartheta_i)}$$

From simple trigonometric considerations we have that:

$$t_{\perp} - r_{\perp} = t_{\parallel} + r_{\parallel} = 1$$

Define the reflection and transmission coefficients as:

$$R = \frac{A_r^2}{A_i^2}$$

$$T = \left(\frac{A_t \cos\vartheta_t}{A_i \cos\vartheta_i} \right)^2$$

We have: R+T=1.
If the angle between the incident beam and the transmitted beam is 90 degrees, then:

$$r_{\parallel} = 0$$

$$\tan\vartheta_i = n$$

This angle is called Brewster's angle and depends on the single material considered.

Diffraction

The phenomenon of diffraction is generated when a wave passes through a slit, causing a series of disparate physical effects.

Fraunhofer's theory is applied at a suitable distance from the slit and reconstructs the optical diffraction figure based on the intensity of the wave, defined as the square of the amplitude.

Calling N the number of slits, b their width, d the distance between the slits, we obtain:

$$u = \frac{\pi b \sin \vartheta}{\lambda}$$

$$v = \frac{\pi d \sin \vartheta}{\lambda}$$

And the angular intensity distribution with respect to the one before the slit is:

$$\frac{I(\vartheta)}{I_0} = \left(\frac{\sin u}{u} \right)^2 \left(\frac{\sin(Nv)}{\sin v} \right)^2$$

Maximum intensity is obtained for

$$d \sin \vartheta = k\lambda$$

This diffraction pattern changes expression if the apertures are circular, rectangular, or otherwise shaped.

In the vicinity of the slit, this theory is not applicable as the hypotheses of geometric optics no longer exist.

For an in-depth study of diffraction at short distances it is necessary to resort to the formalism introduced by the electromagnetic theory of light in the second half of the 19th century.

Diffraction limits the resolution capability of an optical system, as two incident rays before the slit will generate two different diffraction patterns which must be decomposed.

For circular openings, where D is the diameter of the slit, the minimum angle that can be resolved is:

$$\Delta \vartheta_{min} = 1.22 \frac{\lambda}{D}$$

This quantity is crucial for the observation of celestial objects using telescopes and optical telescopes.
For a grating, given a the distance between the peaks of the diffraction pattern and N the number of peaks, the following holds:

$$\Delta \vartheta_{min} = \frac{2\lambda}{Na \cos \vartheta_m}$$

In a geometry composed of N multiple slits, the minimum wavelength spacing that guarantees a separable diffraction pattern is:

$$\frac{\Delta \lambda}{\lambda} = N$$

Interference

On the other hand, interference occurs when two or more light rays add up to give rise to an interference pattern, a typically wave-like phenomenon.
The construction of an elementary interferometer, called Fabry-Perot, leads to modifying the equation relating to the transmission and reflection coefficients by also adding the absorption coefficient: T+R+A=1.
The interference pattern is:

$$\frac{I_t}{I_i} = \left[1 - \frac{A}{1-R}\right]^2 \frac{1}{1 + F \sin^2 \vartheta}$$

$$F = \frac{4R}{(1-R)^2}$$

The fineness of the interferometer is:

$$\Phi = \frac{\pi}{2}\sqrt{F}$$

The maximum frequency resolution is as follows:

$$\Delta f_{min} = \frac{v}{2nd\Phi}$$

Where d is the distance between the two mirrors of the interferometer.

Lenses and mirrors

Geometric optics explains the properties of lenses and mirrors very well. Gauss's formula for lenses can be easily obtained from Fermat's principle and from the geometric approximation according to which, for small angular values:

$$\cos\varphi = 1$$
$$\sin\varphi = \varphi$$

Given v the distance of the object to be displayed, b the distance of the displayed image and R the radius of the spherical surface, we obtain:

$$\frac{n_1}{v} - \frac{n_2}{b} = \frac{n_1 - n_2}{R}$$

If we apply this result to a system composed of a lens having its own refractive index and two radii of curvature in correspondence with the two spherical surfaces and considering that air has an almost unitary refractive index, the above formula becomes:

$$D = \frac{1}{f} = (n_l - 1)\left(\frac{1}{R_2} - \frac{1}{R_1}\right)$$

Where f is the focal length of the lens and D is the dioptric power of the lens.

For a system of two lenses, called d the distance between the lenses, the following holds:

$$\frac{1}{f} = \frac{1}{f_1} + \frac{1}{f_2} - \frac{d}{f_1 f_2}$$

If the lens cannot be considered thin, its thickness also comes into play because double reflections and refractions take place.
In this case, the relationship holds:

$$\frac{1}{f} = \frac{8(n-1)s}{D^2}$$

Where s is the thickness of the lens and D is the diameter.
According to the positive and negative signs of the main quantities, lenses can be classified.
If R is positive we speak of lenses with a concave surface, otherwise the surface is convex.
If f is positive the lens is converging, otherwise it is diverging.
If v is positive we speak of a real object, otherwise of a virtual one (located behind the lens).
If b is positive we have a virtual image, vice versa a real image.
With these conventions it is possible to build lenses to focus on distant or near objects and the main functions of the human eye can be explained.

For mirrors, however, the following relation holds:

$$\frac{1}{f} = \frac{1}{v} + \frac{1}{b} = \frac{2}{R} + \frac{h^2}{2}\left(\frac{1}{R} - \frac{1}{v}\right)^2$$

Where h is the perpendicular distance between the optical axis and the point where the light ray hits the mirror.
A parabolic mirror has no spherical aberration for rays parallel to the optical axis, which is why it is used in telescopes to view light from distant celestial objects.
At the sign convention level, we have the same considerations with respect to what was done for the lenses for R and v, while the opposite convention as regards b and f.

Therefore, lenses and mirrors can be used to zoom in or out on images.
The linear magnification coefficient is:

$$N = -\frac{b}{v}$$

For a telescope it is

$$N = \frac{f_{obiettivo}}{f_{oculare}}$$

Matrix description

A ray of light can be described by a vector with two components determined one by the product between the refractive index and the angle formed between the optical axis and the direction of the ray and the other by the distance from this axis.
Therefore the evolution between a final state and an initial state can be summarized as follows:

$$\begin{bmatrix} n_2\alpha_2 \\ y_2 \end{bmatrix} = M \begin{bmatrix} n_1\alpha_1 \\ y_1 \end{bmatrix}$$

Where M is a matrix having unitary trace and expression of a product of elementary matrices.
The geometrical optics is therefore reduced to the identification of such matrices for the different cases exposed.
For the propagation in free space along a generic distance l, we have:

$$M = \begin{bmatrix} 1 & 0 \\ \dfrac{l}{n} & 1 \end{bmatrix}$$

For refraction on a surface with dioptric power equal to D:

$$M = \begin{bmatrix} 1 & -D \\ 0 & 1 \end{bmatrix}$$

Light also has the property of polarization, i.e. of being more or less intense along particular directions, to the point of being so only in one of them if suitably filtered by an object called a polarizing filter.
This property comes from the electromagnetic wave characteristics of light. Once the minimum and maximum intensities have been defined after the polarizing filter, the polarization is calculated as follows:

$$P = \frac{I_{max} - I_{min}}{I_{max} + I_{min}}$$

With the matrix method it is easy to identify some very simple cases:

Horizontal linear polarizer
$$M = \begin{bmatrix} 1 & 0 \\ 0 & 0 \end{bmatrix}$$

Vertical linear polarizer
$$M = \begin{bmatrix} 0 & 0 \\ 0 & 1 \end{bmatrix}$$

45° linear polarizer$^{\pm}$
$$M = \frac{1}{2} \begin{bmatrix} 1 & \pm 1 \\ \pm 1 & 1 \end{bmatrix}$$

Vertical polarizer $\lambda/4$
$$M = e^{i\frac{\pi}{4}} \begin{bmatrix} 1 & 0 \\ 0 & -i \end{bmatrix}$$

Right circular homogeneous polarizer
$$M = \frac{1}{2} \begin{bmatrix} 1 & i \\ -i & 1 \end{bmatrix}$$

Dispersion

Another property of the optics is the so-called dispersion which manifests itself clearly when light passes through a prism.

The double refraction due to the prism (incoming and outgoing) causes an angular deviation from the original direction of the beam; by varying the angle of incidence of the ray there is a different angular deviation.

In particular, the refractive index of the prism can be defined by considering the minimum value of angular deviation:

$$n = \frac{\sin\left(\frac{1}{2}(\delta_{min} + \alpha)\right)}{\sin\left(\frac{\alpha}{2}\right)}$$

At different wavelengths there are different angular deviations.

This property of prisms was used precisely to separate white light into different colors (we will have an exhaustive explanation of this behavior only after understanding the theory of electromagnetism, in Newton's time and in the following century the scientific explanations on the existence of different colors were somewhat fanciful and devoid of real foundation).

The dispersion of a prism is given by:

$$D = \frac{d\delta}{d\lambda} = \frac{d\delta}{dn}\frac{dn}{d\lambda} = \frac{2\sin\left(\frac{\alpha}{2}\right)}{\cos\left(\frac{1}{2}(\delta_{min} + \alpha)\right)}\frac{dn}{d\lambda}$$

The last term depends on the composition of the prism and is called dispersion coefficient; it is generally negative for visible light.

Optical effects

We can also classify the aberrations of images reflected or transmitted by lenses and mirrors.

Due to the dispersion relationship, chromatic aberration occurs, i.e. the image of an object is reflected or transmitted differently based on the colors that make up the object.

One way to correct this defect is to introduce lenses to compensate for dispersion and focus different wavelengths in the same point.

A spherical lens does not transmit the image perfectly as there are second order effects which we have now neglected; this gives rise to spherical aberration, which is particularly important for focusing distant rays.

That's why telescopes are made up of parabolic mirrors and not spherical ones.

The coma originates from the imperfect planarity of the lenses, in fact far from the optical axis they tend to curve.

Astigmatism originates from a difference in lens thickness and causes elliptical images for all those objects that are not in line with the optical axis.

Finally, distortions are important because they tend to deform the edges of the images of objects and can be corrected by means of suitable combinations of lenses.

In addition, there are special optical effects that become important in particular situations.

Birefringence is due to the fact that the refractive index is not constant along the principal axes.

For a three-dimensional system, for example, it is not certain that the refractive indices along the Cartesian axes are equal to each other.

Dichroism is a consequence of birefringence and occurs when there is a different absorption between the directions parallel and perpendicular to the optical axis.

In this case, optical effects such as double images can be created.

Using particular polarizers it is possible to cause a phase lag between the different directions mentioned above.

Fundamentals of nonlinear optics

Optics, like any other sector of physics, can have non-linear behaviors, the constitutive equations of which differ from those shown previously.

We will only mention the main non-linear optical effects, remembering that they are part of the diffractive optics scheme that we will explain when speaking of electromagnetism.

The Kerr effect occurs when light is immersed in an electric field. A difference in the refractive index is generated in the two directions shown above, proportional to a constant, called Kerr's, which depends on the material considered.

The Pockels effect occurs in crystals that have no center of symmetry, about 20 classes of crystals out of a total of 32. In these crystals a phase lag is generated when an electric field is applied and the polarization changes following an external pressure, thus determining their piezoelectricity.

The Faraday effect is responsible for the rotation of the polarization of light when a magnetic field is applied in the direction of propagation of the light beam.

High school exercises

Exercise 1

Find the frequency range of visible light.

From the wave relation:

$$v = \lambda f$$

We know that light travels with a speed equal to:

$$3.0 \cdot 10^8 \, m/s$$

The wavelengths of visible light range from 400 nm (for violet) to 700 nm (for red) and therefore the frequencies will be:

$$f_1 = \frac{v}{\lambda_1} = \frac{3.0 \cdot 10^8 \, \frac{m}{s}}{4.0 \cdot 10^{-7} \, m} = 7.5 \cdot 10^{14} \, Hz$$

$$f_2 = \frac{v}{\lambda_2} = \frac{3.0 \cdot 10^8 \, \frac{m}{s}}{7.0 \cdot 10^{-7} \, m} = 4.3 \cdot 10^{14} \, Hz$$

Exercise 2

Calculate the critical angle of total reflection for a light beam passing from water to air.

The refractive index of air is 1, that of water 1.33.
Total reflection is obtained when the angle of refraction is 90°.
From Snell's law we have:

$$\frac{sen(i)}{sen(90°)} = \frac{V_{acqua}}{V_{aria}}$$

$$sen(i) = \frac{1}{1,33}$$

$$i = arcsen(0,752) = 48,75°$$

Exercise 3

An object is placed at such a distance from a converging spherical lens that the generated image is twice the size of the object. Knowing that the focal length of the lens is 30 cm, how far away is the object?

Given f the focal length and p the distance, for a converging lens the magnification is given by:

$$G = \frac{f}{f - p}$$

So getting the distance:

$$p = \frac{f \cdot (G - 1)}{G}$$

I our case:

$$p = \frac{30\,cm \cdot (2 - 1)}{2} = 15\,cm$$

The image is therefore virtual.

Exercise 4

An object is placed in front of a converging lens at a distance of 20 cm.
The focal length of the lens is 15cm.
At what distance from the lens does the image form?
What is the magnification factor of the lens?

For a converging lens, the magnification is given by:

$$G = \frac{f}{f - p}$$

$$G = \frac{15\,cm}{15\,cm - 20\,cm} = -3$$

The image is then turned upside down and magnified three times.
Using the law of conjugate points, we have:

$$\frac{1}{p} + \frac{1}{q} = \frac{1}{f}$$

$$\frac{1}{q} = \frac{1}{f} - \frac{1}{p}$$

$$\frac{1}{q} = \frac{p - f}{fp}$$

I our case:

$$q = \frac{fp}{p - f} = \frac{15\,cm \cdot 20\,cm}{20\,cm - 15\,cm} = 60\,cm$$

The image is therefore real.

Exercise 5

A ray of light passes from air to water at an angle of incidence of 45°.
At what angle of refraction does it enter the water?

From Snell's law, knowing the refractive indices of water and air, we have:

$$\frac{sen(r)}{sen(i)} = \frac{n_{aria}}{n_{H_2O}}$$

$$sen(r) = \frac{1,0003}{1,33} \cdot \frac{sqrt(2)}{2} = 0,53182$$

$$r = arcsen(0,53182) = 32,13°$$

Exercise 6

A ray of green light passes through a glass plate with a refractive index of 1.4.
Knowing that the plate is 3 mm thick, how many oscillations does the light make inside it?

Green light has a frequency of:

$$\nu = 6 \cdot 10^{14} \, Hz$$

The speed of light in glass is:

$$V = \frac{c}{n} = \frac{299792458 \, \frac{m}{s}}{1,4} = 214137470 \, \frac{m}{s}$$

The wavelength of green light in glass is:

$$\lambda = \frac{V}{\nu} = \frac{214137470 \, \frac{m}{s}}{6 \cdot 10^{14} \, Hz} = 3,57 \cdot 10^{-7} \, m = 357 \, nm$$

The number of swings made is:

$$n = \frac{d}{\lambda} = 8403$$

University-level exercises

Exercise 1

A ray of white light strikes a glass prism with an aperture angle of 60° with an angle of incidence of 45°.
Due to the scattering of light, the ray separates.
Having said 1.51 the refractive index of the red light and 1.53 that of the violet light, determine the angles of refraction at the point of entry for the two radiations, the angles of incidence at the point of exit for the two radiations, the angles of refraction at the point angle for the two radiations and the angle identified by the two outgoing radiations.

We apply Snell's law to the input:

$$r_R = \arcsin\left(\frac{1}{n_R}\sin i\right) = \arcsin\left(\frac{1}{1.51}\sin 45°\right) = 27.92°$$

$$r_V = \arcsin\left(\frac{1}{n_V}\sin i\right) = \arcsin\left(\frac{1}{1.53}\sin 45°\right) = 27.53°$$

And so since the opening angle is 60°:

$$i'_R = 60° - 27.92° = 32.08°$$

$$i'_V = 60° - 27.53° = 32.47°$$

179

We apply Snell's law to the output:

$$r'_R = \arcsin(n_R \sin i'_R) = \arcsin(1.51 \sin 32.08°) = 53.31°$$

$$r'_V = \arcsin(n_V \sin i'_V) = \arcsin(1.53 \sin 32.47°) = 55.23°$$

From which:

$$\varepsilon = r'_V - r'_R = 55.23° - 53.31° = 1.92°$$

Exercise 2

Given an air-glass diopter with refractive indices 1 en, a concave mirror of radius R is located at a distance R to the right of the diopter.
An object is placed at a positive distance from the diopter.
Find the first image of the object due to the diopter and transverse magnification, the image formed by the mirror and transverse magnification, the second image of the object when the rays come from the right.

The diopter plane equation is:

$$\frac{1}{o_1} + \frac{n}{i_1} = 0$$

From which:

$$i_1 = -no_1 < 0$$

And the magnification is:

$$G_1 = -\frac{1}{n}\frac{i_1}{o_1} = 1$$

For the mirror image we have:

$$o_2 = R - i_1 > 0$$

$$\frac{1}{o_2} + \frac{1}{i_2} = \frac{2}{R}$$

From which:

$$i_2 = \frac{Ro_2}{2o_2 - R} > 0$$

$$G_2 = -\frac{R}{2o_2 - R}$$

If the rays come from the right:

$$o_3 = 2R - i_2 > 0$$

$$\frac{n}{o_3} + \frac{1}{i_3} = 0$$

And then:

$$i_3 = -\frac{o_3}{n} < 0$$

$$G_3 = -\frac{i_3}{o_3} n = 1$$

The final image is virtual, flipped and zoomed out:

$$G = G_1G_2G_3 = -\frac{R}{2o_2 - R} = -\frac{R}{R + 2no_1}$$

Exercise 3

A lens system consists of a first lens with a focal length of 15 cm and a second lens with a focal length of 20 cm placed 70 cm to the right of the first lens.
An object is placed at a distance of 20 cm to the left of the first lens as in the figure:

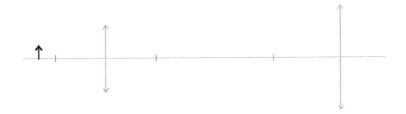

Determine distance and characteristics of the image due to the first lens and that due to the second lens.
Finally, determine the characteristics of the final image.

Applying the lens equation to the first lens, we have:

$$\frac{1}{o_1} + \frac{1}{i_1} = \frac{1}{f_1}$$

From which:

$$\frac{1}{i_1} = \frac{1}{f_1} - \frac{1}{o_1} = \frac{1}{15} - \frac{1}{20} = \frac{4-3}{60} = \frac{1}{60}$$

So the image is real, 60cm away, flipped, and magnified by a factor:

$$G_1 = -\frac{i_1}{o_1} = -\frac{60}{20} = -3$$

For the second lens we have:

$$\frac{1}{o_2} + \frac{1}{i_2} = \frac{1}{f_2}$$

$$o_2 = d - i_1 = 70 - 60 = 10cm$$

$$\frac{1}{i_2} = \frac{1}{f_2} - \frac{1}{o_2} = \frac{1}{20} - \frac{1}{10} = \frac{1-2}{20} = -\frac{1}{20}$$

So the image is virtual, upright, and magnified by:

$$G_2 = -\frac{i_2}{o_2} = -\frac{-20}{10} = +2$$

The final image will be virtual, flipped and zoomed by:

$$G = G_1 G_2 = (-3) \cdot (+2) = -6$$

Exercise 4

Two slits 0.6 mm apart illuminated by a coherent beam of light produce interference on a screen 534 cm away.
Knowing that the fourth maximum of the interference pattern is 2.3 cm from the central peak, determine the wavelength of the incident light.

It is known that the intensity of the interference pattern at very large distances is given by:

$$I = I_0 \cos^2\left(\pi \frac{D}{\lambda}\sin\alpha\right)$$

Therefore, the wavelength is:

$$\lambda = \frac{y_4 D}{4d} = \frac{2.3 \cdot 0.6}{4 \cdot 534} = 646nm$$

Exercise 5

Study the light emitted in the x direction from a coherent source in the following diagram:

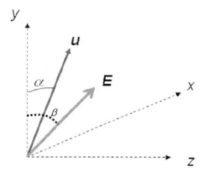

In case of linear or circular polarization.

If the polarization is linear we will have:

$$\vec{E}_{i,L} = E_y \sin(kx - \omega t)\hat{j} + E_z \sin(kx - \omega t)\hat{k}$$

And then:

$$E_y = E_0 \cos\beta$$
$$E_z = E_0 \sin\beta$$
$$\vec{E}_{i,L} = \left(E_0 \cos\beta\hat{j} + E_0 \sin\beta\hat{k}\right)\sin(kx - \omega t)$$

If it's circular:

$$\vec{E}_{t,c} = E_0 \sin(kx - \omega t)\hat{j} + E_0 \cos(kx - \omega t)\hat{k}$$

From which:

$$\vec{E}_{f,c} = (\vec{E}_{t,c} \cdot \hat{u})\hat{u} = \left\{ [E_0 \sin(kx - \omega t)\hat{j} + E_0 \cos(kx - \omega t)\hat{k}] \cdot \hat{u} \right\}\hat{u} =$$
$$= \left\{ E_0 \sin(kx - \omega t)\hat{j} \cdot \hat{u} + E_0 \cos(kx - \omega t)\hat{k} \cdot \hat{u} \right\}\hat{u} =$$
$$= \left\{ E_0 \sin(kx - \omega t)\cos\alpha + E_0 \cos(kx - \omega t)\sin\alpha \right\}\hat{u} =$$
$$= E_0 \sin(kx - \omega t + \alpha)\hat{u}$$

Exercise 6

A screen, illuminated by two monochromatic plane waves of wavelength 700 nm and 420 nm, has two parallel slits 0.5 mm apart.
Find the condition where an interference maximum of the first wave coincides with a maximum of the second wave.
Find the pair of smallest values for which this relationship is satisfied.

The interference pattern at large distances is given by:

$$I_{interf} = I_0 \cos^2\left(\pi \frac{d}{\lambda} \sin\alpha\right)$$

The interference maxima are found when the phase is a multiple of 180° therefore:

$$\pi \frac{d}{\lambda_1} \sin\alpha_1 = n_1 \pi$$

$$\pi \frac{d}{\lambda_2} \sin\alpha_2 = n_2 \pi$$

From which:

185

$$n_1 700 = n_2 420$$

The minimum values are:

$$n_1 = 3,\ n_2 = 5.$$

Exercise 7

An unpolarized wave impinges at an angle equal to that of Brewster on an air-water surface.
Knowing that the refractive index of water is 1.33, calculate the parallel and transverse reflection and transmission coefficients and the global reflection and transmission coefficients.

The Brewster angle is given by:

$$tg\theta_B = n$$

And then:

$$\theta_B = arctg(n) = arctg(1.33) = 53.05°$$

While the transmission angle will be:

$$t = arcsin\left(\frac{\sin\theta_B}{n}\right) = arcsin\left(\frac{0.7993}{1.33}\right) = arcsin(0.6010) = 36.94°$$

The parallel and transverse reflection coefficients are given by:

$$R_{||} = \left(\frac{tg(i-t)}{tg(i+t)}\right)^2 \qquad\qquad R_\perp = \left(\frac{\sin(i-t)}{\sin(i+t)}\right)^2$$

I our case:

$$R_{||} = 0 \qquad\qquad R_\perp = \left(\frac{\sin(53.05-36.94)}{\sin 90}\right)^2 = (\sin 16.11)^2 = 0.0770$$

The parallel and transverse transmission coefficients are:

$$T_\| = 1 - R_\| = 1 \qquad T_\perp = 1 - R_\perp = 1 - 0.0770 = 0.9230$$

While the global ones:

$$I_{rif} = \frac{I}{2}R_\| + \frac{I}{2}R_\perp = I\frac{R_\| + R_\perp}{2}$$

$$R = \frac{I_{rif}}{I} = \frac{R_\| + R_\perp}{2} = \frac{0 + 0.0770}{2} = 0.0385$$

$$T = 1 - R = 1 - 0.0385 = 0.9615$$

Exercise 8

A screen is illuminated with a monochromatic plane wave having a wavelength of 500 nm and has two parallel slits of 0.04 mm width 0.5 mm apart.
Calculate the number of interference maxima included in the central diffraction maximum.

The intensity of diffraction and interference are:

$$I_{diff} = I_0 \left(\frac{\sin(\pi(a/\lambda)\sin\alpha)}{\pi(a/\lambda)\sin\alpha} \right)^2$$

$$I_{interf} = I_0 \cos^2\left(\pi\frac{d}{\lambda}\sin\alpha \right)$$

The central maximum of diffraction is bounded by the first order minima:

$$\pi\frac{a}{\lambda}\sin\alpha_{dif} = \pm\pi \qquad\qquad \sin\alpha_{dif} = \pm\frac{\lambda}{a}$$

Within this interval there are interference maxima such that:

$$\pi \frac{d}{\lambda} \sin \alpha_{int} = \pm k\pi \qquad\qquad \sin \alpha_{int} = \pm k \frac{\lambda}{d}$$

It is therefore necessary that:

$$k \frac{\lambda}{d} \leq \frac{\lambda}{a} \qquad\qquad k \leq \frac{d}{a}$$

From which, indicating the integer part with the square brackets, we have:

$$N = 2\left[\frac{d}{a}\right] + 1 = 2\left[\frac{0.5}{0.04}\right] + 1 = 2[12.5] + 1 = 25$$

Exercise 9

A screen, illuminated with the linear superposition of two plane waves with wavelengths of 600 nm and 450 nm, has a slit of 0.04 mm amplitude. Find a condition where a diffraction minimum of the first wave coincides with a minimum of the second wave.

The diffraction intensity is given by:

$$I_{diff} = I_0 \left(\frac{\sin(\pi(a/\lambda)\sin \alpha)}{\pi(a/\lambda)\sin \alpha} \right)^2$$

Minima occurs when the phase is a multiple of 180° with zero excluded:

$$\pi \frac{a}{\lambda_1} \sin \alpha_1 = n_1 \pi \qquad\qquad \pi \frac{a}{\lambda_2} \sin \alpha_2 = n_2 \pi$$

From which it follows that:

$$n_1 600 = n_2 450$$

The minimum values are:

$$n_1 = 3, \, n_2 = 4.$$

Exercise 10

Two holes of negligible size located at a distance d are illuminated by light of wavelength lambda.
Calculate the diffraction pattern on a very large screen placed at D.

The optical path difference between the two holes is:

$$d \sin \theta$$

And so the phase difference is:

$$\delta = kd \sin \theta$$

The amplitude of the wave is:

$$E \propto 1 + e^{i\delta},$$

And so its intensity is:

$$I \propto |E|^2 \propto \cos^2 \frac{\delta}{2} = \cos^2 \frac{kd \sin \theta}{2} = \cos^2 \frac{\pi d y}{\lambda D}$$

Where is it:

$$k = 2\pi / \lambda$$

The diffraction maxima are obtained for:

$$y_n = n\lambda D/d.$$

Exercise 11

A plane wave of wavelength lambda is incident on a rectangular hole of thickness d.
Calculate the intensity of the diffracted wave immediately after the hole.

In this case we have the so-called Fraunhofer diffraction.
Applying the definition, we have:

$$E \propto \int_{-d/2}^{d/2} dy \, e^{iky\sin\theta} = 2\frac{\sin(kd\sin\theta/2)}{k\sin\theta}, \qquad I \propto |E|^2$$

We note that the function found is the Fourier transform of the rectangle.

Exercise 12

A plane wave of lambda wavelength is incident on a diffraction grating consisting of N small holes of negligible size aligned for a total length d.
Calculate the Fraunhofer diffraction pattern.

We will simply have a summation extended to each hole:

$$E \propto \frac{1}{N}\sum_{n=0}^{N-1} e^{Ikd(n/N)\sin\theta} = \frac{1}{N}\frac{c^N - 1}{c - 1}, \qquad c = e^{ikd/N\sin\theta}$$

$$I \propto |E|^2 \propto \frac{1}{N^2}\frac{\sin^2(kd\sin\theta/2)}{\sin^2(kd\sin\theta/2N)}$$

Two wavelengths are resolved if:

$$\Delta\lambda/\lambda > 1/mN$$

Exercise 13

Two holes of negligible size located at a distance d are illuminated by light of wavelength lambda.
One of the holes is covered by a transparent material of refractive index ne 0.1 mm thick.
The interference fringes with respect to the case of having two open holes have moved by 1 cm.
Calculate the refractive index of the material.

The optical path difference between the two holes is:

$$d \sin \theta$$

And so the phase difference in our case is:

$$\delta \simeq kd\theta + 2\pi s(n-1)/\lambda$$

The amplitude of the wave is:

$$E \propto 1 + e^{i\delta},$$

And so its intensity is:

$$I \propto \cos^2 \left[\frac{\pi}{\lambda} \left(\frac{dy}{D} + s(n-1) \right) \right]$$

Compared to the normal case, the fringes will be moved by:

$$\Delta y = Ds(n-1)/d.$$

Inserting the values you get:

$$n = 1 + \Delta y \cdot d / sD = 1.1.$$

8

WAVES AND OSCILLATORY PHENOMENA

Definition of wave phenomena

Optics confronts us with the fundamental problem of mathematically defining wave phenomena.

In the history of modern physics this need has persisted over the centuries and has had various approaches, first starting from mechanics, later arriving at optics and finally arriving at studies on electromagnetism.

A wave is a periodic perturbation of a given physical entity that transports energy without transporting matter.

Periodicity characterizes every wave phenomenon and we have clearly seen it by introducing wavelength and frequency, which are nothing but the expression of periodicity in spatial and temporal terms, respectively.

In particular, we can define the wavelength as the periodic spatial distance between two particular points, for example the maxima of the wave amplitude (called peaks) or the points where this amplitude is zero (called nodes).

The amplitude instead takes into account the power of the wave.

The fundamental relationship between wavelength and frequency in a material medium is given by:

$$\lambda f = \frac{\lambda_0}{n} f = v = \frac{v_0}{n}$$

Where v is the speed of the wave, zero subscripts account for physical quantities in vacuum, while n is the refractive index.

As can be seen, the speed of a wave depends on the material medium in which it propagates and this dependence is transferred entirely to the wavelength, while the frequency remains constant.

We can define two further quantities related to the previous ones: the wave number and the pulsation.

$$k = \frac{2\pi}{\lambda}$$

$$\omega = 2\pi f$$

Wave equation and solutions

The general equation of a wave was explained in 1747 by D'Alembert:

$$\nabla^2 u - \frac{1}{v^2} \frac{\partial^2 u}{\partial t^2} = 0$$

Being a partial differential equation, it will have different solutions based on the spatial boundary conditions and the temporal initial conditions.

If the wave number and the velocity are parallel to the solution of this equation, then we speak of longitudinal waves and the wave oscillates in the same direction in which it propagates.

If, on the other hand, they are perpendicular, we have the case of transversal waves.

Two different concepts of speed can be defined for a wave.

The phase velocity is given by the ratio between pulsation and wave number, the group velocity by their derivative.

$$v_{ph} = \frac{\omega}{k}$$

$$v_g = \frac{d\omega}{dk} = v_{ph}\left(1 - \frac{k}{n}\frac{dn}{dk}\right)$$

These speeds coincide only if the medium is not dispersive.

Conversely, the group velocity can be less or greater than the phase velocity.

The group speed is the speed with which the energy contribution linked to the wave is transferred and, therefore, characterizes much better the description of a wave phenomenon.

The general solution of the D'Alembert equation in one dimension is given by:

$$u(x,t) = \int_{-\infty}^{+\infty} \left(a(k)e^{i(kx-\omega_1(k)t)} + b(k)e^{i(kx-\omega_2(k)t)} \right) dk$$

This Fourier integral depends both on the boundary conditions and on the dispersion relations.

In Cartesian coordinates, the solutions of D'Alembert's equations are called plane waves.
In one dimension this solution is given by a stationary wave, ie by a wave in which the peaks and nodes correspond to fixed and defined spatial points.
A plane wave is defined by:

$$u(\vec{x},t) = |\vec{u}|\cos\left(\vec{k} \cdot \vec{x} \pm \omega t + \varphi\right)$$

Where φ is the phase, while the sign $\pm$ takes into account the progressive wave and the regressive one (ie its reflection when it arrives at the boundary).
If the boundary condition dictates that the plane wave is zero at the boundary, then the reflected wave will undergo a 90° phase change.
If, on the other hand, the imposition is given by the cancellation of the derivative around the edge, there is no phase change.
An observer in motion will notice a change in frequency of the wave, called Doppler effect, particularly important in acoustics, optics and electromagnetism:

$$\Delta f = \frac{v_f - v_{obs}}{v_f}$$

In spherical coordinates, the wave equation for the radial is given by:

$$\frac{1}{v^2}\frac{\partial^2(ru)}{\partial t^2} - \frac{\partial^2(ru)}{\partial r^2} = 0$$

With a general solution like this:

$$u(r,t) = C_1 \frac{f(r - vt)}{r} + C_2 \frac{g(r + vt)}{r}$$

In cylindrical coordinates, the wave equation for the radial is the Bessel equation:

$$\frac{1}{v^2} \frac{\partial^2 u}{\partial t^2} - \frac{1}{r} \frac{\partial}{\partial r} \left(r \frac{\partial u}{\partial r} \right) = 0$$

The solutions are the Hankel functions. For sufficiently large values of r, the following holds:

$$u(r,t) = \frac{|\bar{u}|}{\sqrt{r}} \cos(k(r \pm vt))$$

For all these equations, the analytical solutions can be calculated only in special cases of symmetry and fairly regular boundary conditions.
In all other cases, it is extremely difficult to determine the analytic solutions also because the general Fourier integral cannot be solved in absolute terms.
To overcome this, it is necessary to resort to numerical solutions through suitable methods of numerical calculation and discretization of physical quantities.
The wave equation describes wave phenomena regardless of their physical nature. Therefore it is possible to study the mechanical waves generated by a vibrating string or by an oscillating membrane or the acoustic waves or the waves of an earthquake or the pressure waves in a gas or the electromagnetic waves.
In addition, propagation mechanisms such as shock waves (for example ultrasonic waves) or solitonic waves (typically those of a tsunami) can also be explained.
Each type of wave just described has typical characteristics which uniquely identify it and which are defined only by probing each individual discipline it belongs to (acoustics, seismics, thermodynamic kinematics, electromagnetism and so on).

An example of a nonlinear wave equation is given by the Kortweg-De Vries relation for wave propagation within a plasma:

$$\frac{\partial u}{\partial t} + \frac{\partial u}{\partial x} - au\frac{\partial u}{\partial x} + b^2\frac{\partial^3 u}{\partial x^3} = 0$$

The third term takes into account the nonlinear effects and the fourth term the dispersion.

Another non-linear equation is that of Van der Pol which takes into account non-linear oscillations:

$$\frac{d^2 x}{dt^2} - \varepsilon\omega_0(1 - \beta x^2)\frac{dx}{dt} + \omega_0^2 x = 0$$

Both equations derive from studies carried out during the twentieth century, therefore well beyond the classical time limit of modern physics.

Waves on a rope

Consider a string stretched along the x-axis and move it slightly from its equilibrium position.

Given a function describing displacement, it satisfies the wave equation.

In fact, once T is defined as the tension of the string and alpha as the angles formed with the x axis (see the figure), we have:

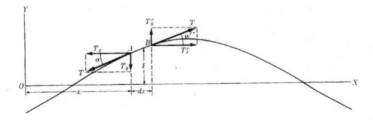

The components of the force are:

$$F_x = T(\cos\alpha' - \cos\alpha)$$
$$F_y = T(\text{sen}\,\alpha' - \text{sen}\,\alpha)$$

Under the assumption of small perturbations, the cosine is approximately one and the sine is confused with the angle itself, therefore:

$$F_x = 0$$

$$F_y = T \frac{\partial \tan \alpha}{\partial x} dx$$

Considering that:

$$\tan \alpha = \frac{\partial \xi}{\partial x}$$

You get:

$$F_y = T \frac{\partial^2 \xi}{\partial x^2} dx$$

Recalling the laws of dynamics and defining the section of the string and its volume density, we have:

$$\frac{\partial^2 \xi}{\partial x^2} = \frac{\rho S}{T} \frac{\partial^2 \xi}{\partial t^2} = \frac{\lambda}{T} \frac{\partial^2 \xi}{\partial t^2}$$

Lambda is defined as the linear mass density.
This equation is completely identical to that of waves with a wave speed equal to:

$$v = \sqrt{\frac{T}{\lambda}}$$

The displacement from the equilibrium position is perpendicular to the direction of propagation (x): we speak of transverse waves.
Note that this condition derives from the fact that the force component in the direction of motion is zero, a consequence of the hypothesis of small perturbations.
If a large displacement is given, waves parallel to the direction of propagation (longitudinal waves) are also generated.

Sound waves in a gas

Let us consider gas contained in a long tube, of unitary section, arranged in the x direction, with a given density and gas pressure at equilibrium.

At the end of the tube we vary the gas pressure through a source (for example a piston) which oscillates periodically; a local variation of pressure and density will be obtained, which will subsequently be transmitted to the rest of the gas.

Assuming small variations, let's consider the generic displacement from the equilibrium position and the mass of gas between two parallel sections as in the figure:

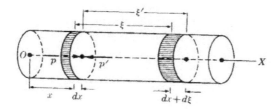

As a consequence of the perturbation, the mass dm undergoes a displacement while the density changes, since the same mass occupies a different volume:

$$dm = (\rho_0 + d\rho)\left(dx + \frac{\partial \xi}{\partial x}dx\right) = \left(\rho_0 + \rho_0 \frac{\partial \xi}{\partial x} + d\rho + d\rho \frac{\partial \xi}{\partial x}\right)dx$$

Neglecting the last term as a higher order infinitesimal, we get:

$$d\rho = \rho - \rho_0 = -\rho_0 \frac{\partial \xi}{\partial x}$$

As far as pressure is concerned, we can find its variation starting from the definition of the compressibility modulus:

$$\beta = -V \frac{dp}{dV}$$

We therefore obtain:

199

$$dp = p - p_0 = \beta \frac{d\rho}{\rho_0} = -\beta \frac{\partial \xi}{\partial x}$$

The pressure variation causes a motion of the gas, as it is numerically equal to the force acting on the mass dm (remember that the section is unitary); therefore the overall force acting on the two sections can be written as:

$$p(x,t) - p(x+dx,t) = -\frac{\partial p}{\partial x} dx = \beta \frac{\partial^2 \xi}{\partial x^2} dx$$

Equating the force to the product of the mass contained between the unit sections and the acceleration, we find:

$$\beta \frac{\partial^2 \xi}{\partial x^2} dx = dm \frac{\partial^2 \xi}{\partial t^2} = \rho_0 dx \frac{\partial^2 \xi}{\partial t^2}$$

or:

$$\frac{\partial^2 \xi}{\partial x^2} - \frac{\rho_0}{\beta} \frac{\partial^2 \xi}{\partial t^2} = 0$$

Thus the displacement of the mass from the equilibrium position satisfies the wave equation.
We deduce that a displacement wave propagates along the gas with speed:

$$v = \sqrt{\frac{\beta}{\rho_0}}$$

If the compressions and expansions are rapid, the gas, schematized as ideal, does not have time to exchange heat with the surrounding environment; for which the wave will propagate in adiabatic conditions.
If we also consider the gas close to equilibrium, we get:

$$-V \frac{dp}{dV} = \gamma p = \beta$$

Therefore, the predicted wave speed is:

$$v = \sqrt{\frac{\gamma p}{\rho_0}}$$

Under standard conditions for air, considered as a diatomic perfect gas, v= 331.61 m/s is found, to be compared with the experimental value 331.45 m/s.
The displacement wave is longitudinal: in fact, the displacement of the mass element occurs along the x axis.
However, it should be noted that each element of mass performs oscillations consisting of compressions and expansions, but there is no net transport of mass along the tube.
A pressure wave and a density wave also propagate along the gas column.
In fact, differentiating with respect to time and with respect to the x coordinate:

$$\frac{\partial^2 p}{\partial x^2} = -\beta \frac{\partial}{\partial x} \frac{\partial^2 \xi}{\partial x^2} = -\rho_0 \frac{\partial}{\partial x} \frac{\partial^2 \xi}{\partial t^2}$$

$$\frac{\partial^2 p}{\partial t^2} = -\beta \frac{\partial^2}{\partial t^2} \frac{\partial \xi}{\partial x} = -\beta \frac{\partial}{\partial x} \frac{\partial^2 \xi}{\partial t^2}$$

Since the derivations with respect to x and with respect to t are independent, their order can be exchanged, from which:

$$\frac{\partial^2 p}{\partial x^2} - \frac{\rho_0}{\beta} \frac{\partial^2 p}{\partial t^2} = 0$$

A pressure wave then propagates along the pipe.
Similarly, we get:

$$\frac{\partial^2 \rho}{\partial x^2} = -\rho_0 \frac{\partial^2}{\partial x^2} \frac{\partial \xi}{\partial x} = -\rho_0 \frac{\partial}{\partial x} \frac{\partial^2 \xi}{\partial x^2}$$

$$\frac{\partial^2 \rho}{\partial t^2} = -\rho_0 \frac{\partial^2}{\partial t^2} \frac{\partial \xi}{\partial x} = -\rho_0 \frac{\partial}{\partial x} \frac{\partial^2 \xi}{\partial t^2}$$

From which:

$$\frac{\partial^2 \rho}{\partial x^2} - \frac{\rho_0}{\beta}\frac{\partial^2 \rho}{\partial t^2} = 0$$

The density wave also propagates with the same speed as the displacement wave.

harmonic waves

We study a particular type of plane wave, the harmonic wave which can be expressed in the following two equivalent forms:

$$\xi(x,t) = \xi_0 \, \text{sen} \, k(x - vt)$$
$$\xi(x,t) = \xi_0 \cos k(x - vt)$$

The multiplying factor is called wave amplitude; the constant k is necessary for dimensional reasons (the argument of a trigonometric function cannot have dimensions), is the inverse of a length and is called the wave number.
It is usually written, inserting the wave number inside the brackets:

$$\xi(x,t) = \xi_0 \, \text{sen}(kx - \omega t)$$
$$\xi(x,t) = \xi_0 \cos(kx - \omega t)$$

and the size:

$$\omega = kv$$

is called the pulsation of the harmonic wave.
For convenience, we reason only in terms of the sine function.
If we fix an instant of time and 'take a picture' of the function, we see that it is a sinusoid along the x axis.
The spatial period (or wavelength) is given by:

$$\lambda = x_2 - x_1 = \frac{2\pi}{k}$$

While the time period is given by:

$$T = t_2 - t_1 = \frac{2\pi}{\omega}$$

Recalling the definition of frequency, we have the fundamental relation of waves which links frequency, wavelength and wave speed:

$$\lambda v = v$$

The argument of the sine function is called the phase of the wave; in more general terms it can be written as:

$$kx - \omega t + \phi$$

Fourier series decomposition

Given a continuous function with a single value x = x(t), periodic of period T.
Fourier's theorem ensures that it can always be represented by a sum of infinite sinusoidal (harmonic) terms having frequencies multiples of that of the given function.
We therefore have:

$$x(t) = a_0 + a_1 \cos \omega t + a_2 \cos 2\omega t + ... + a_k \cos k\omega t + ... + b_1 \operatorname{sen} \omega t + ... + b_k \operatorname{sen} k\omega t + ...$$

The various coefficients a and b are calculated as follows.
We integrate both sides of the previous relation between 0 and T.
All the integrals of the sine and cosine functions are zero and it results:

$$a_0 = \frac{1}{T} \int_0^T x(t)\,dt$$

Therefore, the first coefficient of the Fourier series expansion of x(t) is equal to the mean value of the function x(t) in an interval equal to the period.
Moreover:

$$a_p = \frac{2}{T} \int_0^T x(t) \cos p\omega t \, dt$$

$$b_p = \frac{2}{T} \int_0^T x(t) \operatorname{sen} p\omega t \, dt$$

We observe that non-periodic functions (such as those representing an impulsive phenomenon) can be adequately represented by a sum of infinite harmonic terms.

$$x(t) = \int_0^\infty \left[a(\omega) \operatorname{sen} \omega t + b(\omega) \cos \omega t \right] d\omega$$

In this case we speak of a Fourier integral.
We have:

$$a(\omega) = \frac{1}{\pi} \int_{-\infty}^{+\infty} x(t) \operatorname{sen}(\omega t) \, dt$$

$$b(\omega) = \frac{1}{\pi} \int_{-\infty}^{+\infty} x(t) \cos(\omega t) \, dt$$

The considerations made regarding the Fourier series (or integral) should clarify the importance assumed in the treatment of harmonic waves: in fact, through them, any wave phenomenon, periodic or impulsive, can be 'decomposed' into elementary components, the harmonic waves to be precise.
For this reason, from now on we will refer almost exclusively to harmonic waves.

Spherical waves

So far we have considered waves that propagate along the x-axis and have constant amplitude (plane waves).

Their characteristic is that the phase of the wave, at a given instant, is constant at all points x = constant, which represents in space a plane orthogonal to the x axis.

That is why such waves are called planes.

In general we define a wavefront as a surface on which the phase of the wave is constant at a given instant.

The wave front moves with the speed of wave propagation.

The line orthogonal to the wave front in a given point represents, in that point, the direction of propagation of the wave and of the energy associated with it; this line is called a radius.

Plane waves are not the only waves we can conceive of.

Let us imagine that a point source placed at the origin generates waves that propagate in all directions (isotropic emission); by symmetry the wavefronts of these perturbations are spherical.

In this case we speak of spherical waves; they are also possible solutions of the three-dimensional D'Alembert equation.

Let's try to obtain some characteristics of these waves, without solving the equation.

The intensity of a wave is proportional to the square of the amplitude:

$$I = \frac{1}{2}\rho_0 \omega^2 v \xi(r)^2$$

where we will now assume that the amplitude can be a function of the distance r from the source.

The average power transmitted through a spherical surface of radius r is therefore:

$$P_m = IS = I 4\pi r^2 = \frac{1}{2}\rho_0 \omega^2 v \xi(r)^2 4\pi r^2$$

This power must remain constant, whatever the spherical surface, because it corresponds to the average power emitted by the source.

This implies that for a spherical wave the amplitude decreases with distance:

$$\xi(r) = \frac{\xi_0}{r}$$

If the spherical wave is harmonic we will write:

$$\xi(r,t) = \frac{\xi_0}{r}\,\text{sen}(kr - \omega t)$$

Obviously, the intensity of the wave varies inversely proportional to the square of the distance from the source.

We have drawn these conclusions using the principle of conservation of energy: the power emitted by the source at a given instant we must find it again at a subsequent instant on a spherical surface of suitable radius.

We observe that as the distance from the source increases, the (spherical) wavefronts have an increasing radius, so that a limited portion of the spherical wavefront can be approximated by a flat surface.

The plane wave is therefore an approximation of the spherical wave, when placed at a sufficient distance from the source.

In this case the amplitude of the spherical wave will not vary much over limited distances and can be considered constant, as required for a plane wave.

Wave properties

If we take up the expression of the pressure in a gas crossed by a wave, called S the section of the pipe, the force F on an element of the column is:

$$(p - p_0)S = -\beta S \frac{\partial \xi}{\partial x}$$

The instantaneous power is:

$$P = \beta S \xi_0^2 k \omega \cos^2(kx - \omega t)$$

For the average power it is necessary to average the previous expression over a period:

$$P_m = \frac{1}{2} \rho_0 \omega^2 S v \xi_0^2$$

If we divide by the section S we obtain the intensity I, which represents the average energy that passes through a unitary section orthogonal to the direction of propagation in the unit of time (remember that power is the energy per unit of time):

$$I = \frac{1}{2} \rho_0 \omega^2 v \xi_0^2$$

The intensity can also be written as:

$$I = \frac{(\Delta p)^2_{max}}{2 \rho_0 v}$$

Sound waves extend by convention over a frequency range between 20 and 20,000 Hz; in fact, all the sounds audible to the human ear fall within this interval.
Waves whose frequencies extend beyond the limit of hearing (> 20,000 Hz) are called ultrasound. The frequencies of ultrasound correspond to rather small wavelengths: for example, there is 1.7 cm in air and 7.5 cm in water. This allows to form thin and well collimated ultrasound beams, useful in numerous technical applications, such as for the study of defects in metallic and polymeric materials, for cleaning treatments of the surfaces of materials, for the measurement of distances in conditions of non-visibility (sonar), for diagnostic investigations (ultrasounds).

The speeches just made presuppose that there is no absorption by the medium in which the wave propagates; in practice, phenomena of internal friction always involve a certain absorption.
Experimentally we find an exponential decrease of the intensity:

$$I(x) = I_0 \exp(-\gamma x)$$

being gamma a coefficient (absorption coefficient) which depends on the medium, having as dimension the inverse of a length.

Let us consider two harmonic waves propagating in the same direction and in the same direction, emitted by two sources placed at different points on the x-axis and suppose for the moment that they have the same frequency and the same amplitude, thus being able to differ at most in phase initial. Without affecting the generality we can think that one source is located at the origin, while the other is located at a distance d from the first.
The resulting perturbation (called interference between the two waves) is:

$$\xi = \xi_1 + \xi_2 = 2\xi_0 \cos[kd - \frac{\phi}{2}] \, \text{sen}[k(\frac{2x - d}{2}) - \omega t + \frac{\phi}{2})$$

The resulting wave is still sinusoidal and its amplitude depends on the initial phase difference and the distance between the sources.
If the two sources emit with the same initial phase (they are said to be synchronous).
For given values of k they are in phase, i.e. they interfere constructively and the oscillation will have maximum amplitude, while for other values of k they are in phase opposition and the amplitude of the oscillation is always zero. This is called destructive interference.
For distances that are multiples of the wavelength, there are interference maxima, while for distances that are odd multiples of half a wavelength, there are interference minima.

A well-known effect linked to the overlapping of sounds having frequencies that are not very different from each other is that of beats, consisting of an alternation over time of constructive and destructive interference perceived as an increase and subsequent decrease in the intensity of the sound with a repetition frequency equal to the difference between the two frequencies of the component sounds.
Given two generic waves:

$$\xi_1 = \xi_0 \operatorname{sen}(k_1 x - \omega_1 t)$$
$$\xi_2 = \xi_0 \operatorname{sen}(k_2 x - \omega_2 t)$$

It is clear that in certain instants the two waves will be in phase, in others they will be in phase opposition.
The resulting wave is:

$$\xi = \xi_1 + \xi_2 = 2\xi_0 \cos\left[\frac{k_1 - k_2}{2} x - \frac{\omega_1 - \omega_2}{2} t\right] \operatorname{sen}\left[\frac{k_1 + k_2}{2} x - \frac{\omega_1 + \omega_2}{2} t\right]$$

This wave can be rewritten like this:

$$\xi = \xi_1 + \xi_2 = 2\xi_0 \cos\left[\frac{k_1 - k_2}{2} x - \frac{\omega_1 - \omega_2}{2} t\right] \operatorname{sen}\left[kx - \omega t\right]$$

which is a harmonic wave (sine factor) with frequency and wavelength equal to the arithmetic mean of the corresponding magnitudes of the component waves; this wave is called carrier.
However, its amplitude is not constant but varies over time (modulating wave) with a frequency equal to the semi-difference of the frequencies of the components.

The phenomenon of beats is useful for determining an unknown frequency by comparing it with a source capable of emitting at known frequencies: the frequency is varied until the beats are obtained and the measurement of the frequency of the modulator leads back to the unknown frequency.

The absolute tuning of the instruments is produced by eliminating the beats generated by the A played simultaneously by the instrument and by a standard tuning fork at 440 Hz (or equivalently to an electronic oscillator of the same frequency).

Suppose a traveling plane wave propagating in a medium encounters an obstacle which constitutes a large discontinuity with respect to the wavelength.

If we indicate with i and r the angles that a ray (i.e. the normal to the wave front) forms with the normal to the wall at the point of incidence of the wave, the following two laws hold:

1) the incident ray, the reflected ray and the normal to the reflecting surface at the point of incidence all lie in the same plane.
2) the angle of incidence is equal to that of reflection (i = r).

These laws, established experimentally by Snell, can be deduced from general principles (Huygens principle).

They also apply when the reflecting surface is not flat, and for waves of a different nature from those considered here.

We now want to determine the phase relationship between the incident and reflected waves.

We first consider a rigid reflecting surface at the origin.

The incident wave is harmonic.

The stiffness constraint imposes the condition that the overall displacement is zero on the wall:

It can be seen that the reflected wave has the opposite sign to the incident wave.

In the case of longitudinal waves:

$$\frac{\partial \xi_i}{\partial x} = \xi_0 k \cos(kx - \omega t + \phi_i)$$

$$\frac{\partial \xi_r}{\partial x} = \xi_0 k \cos(kx + \omega t + \phi_r)$$

From this it can be deduced that the pressure variations of the incident and reflected waves are equal on the wall.

As an example we can consider the case of a sound wave that propagates in the air contained in a rigid-walled pipe and terminated by a wall that is also rigid.

The case of a wave on a string having one end fixed to a rigid wall is analogous, although the wave in this case is transverse.

In the study of the reflection of sound waves, the results we have obtained are valid, as already mentioned, in the hypothesis that the wave encounters large obstacles compared to the wavelength of the incident sound.

If the size of the obstacles is comparable with the wavelength, diffraction phenomena occur and the laws of reflection no longer apply.

An example is the passage of a spherical elastic wave through an opening made on a circular obstacle that stands in its way; if the aperture is large compared to the wavelength, the perturbation that propagates beyond the aperture involves a limited region, with an acoustic 'shadow' region, analogously to what happens in optics if a brush of light strikes a large aperture relative to the wavelength.

However, if the opening is narrowed, it is observed that the area of acoustic shadow narrows, until it tends to disappear when the opening has dimensions of the order of the wavelength of the sound.

Obviously the sound intensity is not the same at all points.

A simple explanation of the phenomenon can be found by assuming that the points of the obstacle hit by the wave front in turn become sources of secondary spherical waves; the interference of the latter in the areas beyond the opening produces the sound disturbance.

Now, if the aperture is large, the area affected by constructive interference practically coincides with the geometric projection of the sound source through the surface; other regions of space experience destructive interference, and sound is practically not present in them.

If the opening becomes narrower, the secondary spherical waves produce constructive interference (and therefore a sound perturbation) in ever wider regions of the previous acoustic gray area up to invading, in the limit, the whole region, albeit, as mentioned, with different intensity at different points.

The phenomenon of diffraction also occurs when an obstacle is placed in the path of the sound wave; it is a dual case of the previous one.

With purely geometrical considerations it would be concluded that the sound is not perceptible in the region where the obstacle casts its shadow, assuming to replace the sound source with a luminous one.

However, it is noted that sound also propagates in this area, albeit with reduced intensity.

An elementary example is constituted by the fact that each of us is able to hear, in the open air, his own voice while speaking; this means that the sound disturbance reaches our ears, which are also placed in the 'shadow area' with respect to the source (the mouth).

As another example we cite the fact of being able to hear a person speaking with his back turned towards us.

The results obtained so far assume that the source (S) and the observer-receiver (R) are mutually at rest.
If there is a relative motion between them, the frequency of the sound measured by R is different from that emitted by S. This is the Doppler effect.
We define the speeds of the source and receiver with respect to the medium supporting the sound perturbation and the speed of the sound wave in the medium.
We have that the time interval between two wavefronts is given by:

$$t_2 = t_1 + \frac{V - V_S \cos \alpha_S}{V + V_R \cos \alpha_R} T$$

The perceived frequency is therefore different given by:

$$\nu' = \frac{V + V_R}{V - V_S} \nu$$

Note that, in general, the frequency variation is different, with the same relative speed between S and R, depending on whether it is S or R that moves.
Finally, note that, if the displacement takes place in a direction orthogonal to the line joining S – R, the Doppler effect is zero.

Standing waves

Let us now study a particular but very important case of interference, due to elastic waves propagating in the same direction but in opposite directions.
Suppose the two waves have the same amplitude:

$$\xi_1 = \xi_0 \, \text{sen}(kx - \omega t)$$
$$\xi_2 = \xi_0 \, \text{sen}(kx + \omega t)$$

The resulting wave:

$$\xi = \xi_1 + \xi_2 = \xi_0 [\text{sen}(kx - \omega t) + \text{sen}(kx + \omega t)] = 2\xi_0 \, \text{sen}(kx) \cos(\omega t)$$

This function cannot represent a wave, as spatial and temporal variables are separate.
If we consider a particular point, it performs sinusoidal oscillations with the frequency of the component waves; however, the amplitude of oscillation depends on the choice of point.
In particular for

$$kx = (2n+1)\frac{\pi}{2}$$

the amplitude of oscillation is maximum: the values of x for which this situation occurs are called ventri.
At points where:

$$kx = n\pi$$

the amplitude of oscillation is always zero: they are the nodes.
An oscillation belly and the adjacent node are separated by a distance given by:

$$k\Delta x = \frac{\pi}{2}$$

That is:

$$\Delta x = \frac{\lambda}{4}$$

Two successive nodes (or two bellies) are separated by a double distance.
Observe that, after standing waves are established in a region, there is no transfer of energy from one point to another.
The fact is evident if one bears in mind that energy cannot flow through the nodes, since they are fixed points.
Strictly speaking therefore, a standing wave does not propagate energy.
It is the denomination itself of 'wave' which in this case is equivocal; it is called this only because it derives from the superimposition of two waves of equal frequency and amplitude, one progressive and one regressive.
The simplest way to obtain stationary waves is to exploit the reflections of a wave that occur in the presence of a discontinuity of the propagation

medium (rigid wall): we will therefore have stationary waves on a string fixed at the two ends or in a tube containing air, closed at the ends.

Plucking a string taut in one point we generate, as we have seen, waves which, in a short transient, propagate ending up at the fixed ends of the string.

Here they must be reflected by inverting the phase: the string becomes the seat of progressive and regressive waves which interfere.

The only modes that 'survive' are the standing waves, also called the normal modes of vibration of the string.

Their characteristics depend on the length L of the string: if we require that the displacement of the wave at the two ends is zero, we have the following condition:

$$\lambda = \frac{2L}{n}$$

with n positive integer.

If this condition is not satisfied, the interference between the various reflected waves is destructive and the vibration ceases very quickly.

The wavelengths of the normal modes cannot therefore be chosen arbitrarily, but are related to the length of the string:

$$\lambda = 2L, L, \frac{2}{3}L, \frac{L}{2}, \ldots$$

That is, the modes are quantized.

The mode having the maximum wavelength has the nodes in the extremes themselves (fundamental or first harmonic); the second way (second harmonic) has a knot in the middle of the string, the third way has two knots, and so on.

The general wave that arises on a string with fixed ends is a linear combination of normal modes.

The same goes for a tube closed at the ends, as in many wind instruments.

A vibrating string taut in space transmits the vibration to the surrounding air molecules, generating sounds of very small amplitude: it is therefore not an efficient source.

To increase the efficiency of bowed musical instruments, the strings are therefore stretched near cavities which, by resonating, can reinforce the vibrations produced in the air directly by the strings.

In fact, in this case it is a question of exciting the cavity in some way, obtaining a real resonance: the sound comes out strongly amplified.

This is obtained by providing the cavity with suitably arranged and shaped openings.

The cavity can be given different shapes: experience shows that the fundamental frequency of the cavity increases with the surface area, decreases with its volume and practically does not depend on the shape.

Oscillatory phenomena

Any oscillatory phenomenon can be described by an infinite sum of single elementary oscillatory motions called harmonics which satisfy the following harmonic equation:

$$\nabla^2 f = 0$$

The infinite sum is generally given by a Fourier series and the single contributions are called main harmonic, second harmonic and so on based on the value of the oscillation pulse.
A general form of harmonic, solution of the harmonic equation, is given by:

$$f(t) = |f|e^{i(\omega t \pm \varphi)} = |f|\cos(\omega t \pm \varphi)$$

For harmonic oscillations, the following integration and derivation rules apply:

$$\int f(t)dt = \frac{f(t)}{i\omega}$$

$$\frac{d^n x(t)}{dt^n} = (i\omega)^n x(t)$$

Oscillatory phenomena describe many physical situations, both mechanical and electrical.
Mechanically, a spring with constant C, damping k and mass m, to which a harmonic force is applied, has an equation of motion deriving from Newtonian dynamics as follows:

$$m\ddot{x} = F(t) - k\dot{x} - Cx$$

Whose solutions of the motion are given by:

$$x = \frac{F}{m\left(\omega_0^2 - \omega^2\right) + ik\omega}$$

$$\dot{x} = \frac{F}{k + i\delta\sqrt{CM}}$$

$$\omega_0 = \sqrt{\frac{C}{m}}$$

$$\delta = \frac{\omega}{\omega_0} - \frac{\omega_0}{\omega}$$

The system impedance and the system quality factor are given by:

$$Z = \frac{F}{\dot{x}}$$

$$Q = \frac{\sqrt{Cm}}{k}$$

The frequency with the minimum value of the impedance modulus is called speed resonant frequency and coincides with ω_0.

The resonant frequency in amplitude is instead determined by the minimum value of $i\omega Z$.

It is given by

$$\omega_A = \omega_0 \sqrt{1 - \frac{Q^2}{2}}$$

The damping frequency is

$$\omega_D = \omega_0 \sqrt{1 - \frac{1}{4Q^2}}$$

For an oscillation having small dampings, the time for which the perturbed system returns to the initial state of rest is:

$$T_D = \frac{2\pi}{\omega_D}$$

For oscillations with strong damping we have instead:

$$x(t) \approx x_0 \exp(-\frac{t}{\tau})$$

Finally, for an oscillation at critical damping we have:

$$k^2 = 4mC$$
$$\omega_D = 0$$

The oscillatory phenomena described by harmonic motions represent very well the kinematics and dynamics of the pendulum and have been studied since Newton's time for this very reason.
An oscillatory phenomenon is characterized by a period which is the inverse of a frequency.

For electrical phenomena, impedance is the sum of a real and an imaginary part, while the phase angle is defined as the arctangent of the ratio between the imaginary and real parts:

$$Z = R + iX$$
$$\varphi = \arctan\left(\frac{X}{R}\right)$$

The impedance of a resistor is simply equal to R, that of an inductance is $i\omega L$, that of a capacitance $\frac{1}{i\omega C}$. The quality factor is

$$Q = \omega \frac{L}{R}$$

Define the absolute impedances and pulsations as:

$$Z_0 = \sqrt{\frac{L}{C}}$$

$$\omega_0 = \frac{1}{\sqrt{LC}}$$

you can derive these rules for connections in series or in parallel:

Series:

$$Z_{tot} = \sum_i Z_i$$

$$L_{tot} = \sum_i L_i$$

$$\frac{1}{C_{tot}} = \sum_i \frac{1}{C_i}$$

$$Q = \frac{Z_0}{R}$$

$$Z = R(1 + iQ\delta)$$

Parallel:

$$\frac{1}{Z_{tot}} = \sum_i \frac{1}{Z_i}$$

$$\frac{1}{L_{tot}} = \sum_i \frac{1}{L_i}$$

$$C_{tot} = \sum_i C_i$$

$$Q = \frac{R}{Z_0}$$

$$Z = \frac{R}{(1 + iQ\delta)}$$

High school exercises

Exercise 1

Given a wave phenomenon of 14 waves per minute and with a wavelength of 34 meters, find the propagation speed of the phenomenon.

The frequency of the phenomenon is:

$$f = \frac{14}{60} = 0.23\,Hz$$

The speed of propagation is given by:

$$v = \lambda f = 34\,m \times 0.23\,s^{-1} = 7,8\,\frac{m}{s}$$

Exercise 2

A wave with a frequency of 4.5 Hertz has an amplitude of 12 cm and a wavelength of 27 cm.
Calculate the distance traveled by a wave crest in 0.5 seconds.

In one second there will be 2.25 complete oscillations and therefore the distance covered by a crest will be:

$$\Delta s = 2.25\,s^{-1} \cdot 0.27\,m = 0.61\,m$$

Exercise 3

A wave has a speed of 240 meters per second and a wavelength of 3.2 meters.
Calculate the frequency and period of the wave.

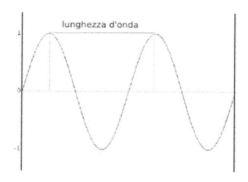

The speed is given by:

$$v = \frac{\lambda}{T}$$

And so the period is:

$$T = \frac{\lambda}{v} = \frac{3.2\,m}{240\,\frac{m}{s}} = 0.013\,s$$

The frequency is the inverse of this number:

$$f = \frac{1}{T} = \frac{1}{0.013\,s} = 75\,Hz$$

Exercise 4

Write the equation for a one-dimensional wave moving in the negative x-direction, having an amplitude of 0.01 meters, a frequency of 550 Hertz, and a velocity of 330 meters per second.

The wave equation is given by:

$$y(x,t) = A\sin(kx - \omega t)$$

Where is it:

$$\omega = \frac{2\pi}{T}$$

$$k = \frac{2\pi}{\lambda}.$$

In our case we have:

$$\omega = 2\pi f = 2\pi \times 550$$

$$\lambda = \frac{v}{f} = \frac{330}{550} = 0.6$$

$$k = \frac{2\pi}{0.6}$$

So the equation will be:

$$y\left(x, t\right) = 0.010 \sin \left(\frac{2\pi}{0.6}x - 2\pi \times 550t\right)$$

Exercise 5

Write the equation for a transverse sinusoidal wave moving in the positive y-direction with wavenumber 60, period 0.2 seconds, and amplitude 3 mm.

The equation is given by:

$$z\left(y, t\right) = A \sin \left(\frac{2\pi}{\lambda}y - \frac{2\pi}{T}t\right)$$

Substituting the numerical values, we have:

$$z = 3.0 \cdot 10^{-3} \sin \left(3.77y - 31.4t\right)$$

Exercise 6

A string has a linear density of 7.2 grams per meter and is subjected to a tension of 150 N.
The length of the rope is 90cm and is held firm at the ends.
Calculate the speed, wavelength and frequency of the wave phenomenon.

This is a standing wave according to the scheme:

The speed is given by:

$$v = \sqrt{\frac{T}{\mu}} = \sqrt{\frac{150\,N}{7.2 \cdot 10^{-3}\,\frac{kg}{m}}} = 144\,\frac{m}{s}$$

Since there are 3 oscillations along the length of the string we have:

$$\lambda = \frac{2L}{3} = \frac{2 \times 0.90\,m}{3} = 0.6\,m$$

And so the frequency is:

$$f = \frac{v}{\lambda} = \frac{v}{2L}n = \frac{144\,\frac{m}{s} \times 3}{1.80\,m} = 240\,Hz$$

Exercise 7

The lowest note on a piano is an A which is four octaves below the A of frequency 440 Hz.

The highest note is a C, four octaves above middle C at 261.7 Hz.
Calculate frequencies and wavelengths of these sounds.

There are stationary waves with frequency:

$$f = \frac{v}{2L}$$

The lowest will have frequency:

$$\frac{f_{La_{basso}}}{f_{La}} = \frac{\frac{v}{16L}}{\frac{v}{2L}} = \frac{1}{8}$$

$$f_{La_{basso}} = \frac{440}{4} = 27.5\,Hz$$

While the highest do:

$$f_{Do_{alto}} = 16f_{Do} = 4187\,Hz$$

The respective wavelengths will be:

$$\lambda_{La_{basso}} = \frac{343}{27.5} = 12.5\,m$$

$$\lambda_{Do_{alto}} = \frac{343}{4187} = 8.2\,cm$$

Exercise 8

A 120 cm long rope is stretched between two fixed supports.
Find the three maximum wavelengths for the standing waves on this string.

The wavelengths of a standing wave are given by:

$$\lambda = \frac{2L}{n}$$

Being n in the denominator, the maximum values are obtained by the smallest n, i.e.:

$$\lambda_1 = 2L(n=1) = 2.40\,m$$
$$\lambda_2 = \frac{2L}{n}(n=2) = 1.20\,m$$
$$\lambda_3 = \frac{2L}{n}(n=3) = 0.80\,m$$

Exercise 9

A metal tube of length L is struck at one end.
At the end of the tube, two sounds are heard, one that has traveled inside the tube and one that has traveled through the air.
Given v the speed of sound, find the time interval between the two sounds.
Assuming t=1 second and the steel pipe, find L.

The time difference is given by:

$$\Delta t = \frac{L}{v} - \frac{L}{V} = \frac{L\,(V-v)}{vV}$$

The speed of sound in air is 331 meters per second, that of steel 5,941 meters per second.
If the time interval is one second, the tube will have length:

$$L = \frac{\Delta t v V}{(V-v)} = \frac{1\,s \times 5491\,\frac{m}{s} \times 331\,\frac{m}{s}}{(5491-331)\,\frac{m}{s}} = 352\,m$$

Exercise 10

A sound wave of length 40 cm enters the following tube:

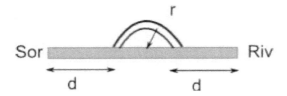

Find the smallest radius r such that a sound is heard at the detector.

An interference is generated between the wave that follows the horizontal path and the one that travels along the curvilinear section.
The path difference is given by:

$$l_1 = 2s + 2r$$
$$l_2 = 2s + \pi r$$
$$\Delta l = r(\pi - 2)$$

So the phase difference is:

$$\phi = \frac{\Delta L}{\lambda} 2\pi = \frac{r(\pi - 2)}{0.40\ m} \cdot 2\pi$$

The detector minimum is obtained for:

$$\dot\phi = \left(m + \tfrac{1}{2}\right) 2\pi$$

For m=0 we have:

$$\frac{r(\pi - 2)}{\frac{4}{10}} \cdot 2\pi = \pi$$

And then:

$$r = \frac{1}{5\,(\pi - 2)} = 0.175\,m = 17.5\,cm$$

University-level exercises

Exercise 1

Given a string welded to another at the point x=0 with a uniform tension but with non-uniform linear mass, let there be an incident wave which gives rise to a reflected and a transmitted wave.
Find the formulation of these waves under the assumption that:

$$f(0^-, t) = f(0^+, t) \, ,$$

$$\left. \frac{\partial f}{\partial x} \right|_{0-} = \left. \frac{\partial f}{\partial x} \right|_{0+}$$

We name the waves in this way, respectively the incident one, the transmitted one and the reflected one.

$$g_I(x, v_1 t)$$
$$g_T(x - v_2 t)$$
$$h_R(x + v_1 t)$$

Where we have already assumed that the reflected wave has the same speed as the incident one.
We have:

$$g_I(-v_1 t) + h_R(v_1 t) = g_T(-v_2 t) \, ,$$
$$\frac{\partial g_I}{\partial x}(-v_1 t) + \frac{\partial h_R}{\partial x}(v_1 t) = \frac{\partial g_T}{\partial x}(-v_2 t) \quad \text{per } x = 0 \, ;$$

Place:

$$\xi \equiv x - v_1 t \,,$$

You get:

$$\frac{\partial g}{\partial x} = \frac{\partial g}{\partial \xi} \cdot \frac{\partial \xi}{\partial x}$$

$$\frac{\partial g}{\partial t} = \frac{\partial g}{\partial \xi} \cdot \frac{\partial \xi}{\partial t}$$

From which:

$$\frac{\partial g}{\partial x} = 1$$

$$\frac{\partial g}{\partial t} = -v_1 \,, \quad +v_1 \,, \quad -v_2 \,,$$

Based on the three functions chosen.
Therefore:

$$\frac{\partial g_I(-v_1 t)}{\partial t} \left(-\frac{1}{v_1} \right) + \frac{\partial h_R(v_1 t)}{\partial t} \left(+\frac{1}{v_1} \right) = \frac{\partial g_T(-v_2 t)}{\partial t} \left(-\frac{1}{v_2} \right)$$

Solving:

$$g_T(x - v_2 t) = \frac{2 v_2}{v_1 + v_2} g_I\left(\frac{v_1}{v_2} x - v_1 t \right) + costante \,.$$

The reflected wave is calculated with a similar procedure.

$$h_R(x + v_1 t) = \frac{v_2 - v_1}{v_1 + v_2} \cdot g_I(-x - v_1 t) + C \,.$$

Exercise 2

Four identical sources arranged as:

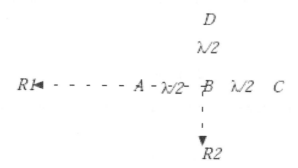

They emit waves of a specific lambda wavelength.
In the figure, R are the receivers located at much greater distances.
Calculate the ratio between the powers received by the two reflectors.
What happens if B is turned off? What if it's D?

The power is given by:

$$I_E \propto |E_A + E_B + E_C + E_D|^2:$$

And so we have:

$$I_1 \quad \propto \quad |e^{ik(r-\lambda/2)} + e^{ikr} + e^{ik(r+\lambda/2)} + e^{ik\sqrt{r^2+\lambda^2/4}}|^2 \propto |-1+1-1+1|^2$$
$$I_2 \quad \propto \quad |e^{ik\sqrt{r^2+\lambda^2/4}} + e^{ikr} + e^{ik\sqrt{r^2+\lambda^2/4}} + e^{ik(r+\lambda/2)}|^2 \propto |1+1+1-1|^2$$

The report is:

$$I_1/I_2 = 0/4$$

If B is turned off:

$$I_1/I_2 = 1/1.$$

If it is turned off D:

$$I_1/I_2 = 1/9.$$

So the first receiver will never be able to tell whether D or B is off, while the second can.

Exercise 3

Two strings with different linear densities are connected, in particular the second one has:

$$\mu' = n^2\mu.$$

What happens when a transverse wave arrives at the junction of the two strings?

Given y(x,t) the distortion generated by the wave, the equation of the mode is:

$$ma = F \qquad : \qquad \mu\frac{\partial^2 y}{\partial t^2} = \tau\frac{\partial^2 y}{\partial x^2}$$

The tension of the string is always the same while the speed changes at the juncture point:

$$v = \sqrt{\tau/\mu(x)} = \omega/k$$

The incident wave will have a reflected and a transmitted component:

$$y = a \begin{cases} e^{i(kx-\omega t)} + R\, e^{i(-kx-\omega t)} & \text{per } x < 0 \\ T e^{i(k'x-\omega t)} & \text{per } x > 0 \end{cases}$$

By imposing continuity on the conjunction we have:

$$T = \frac{2k}{k+k'} = \frac{2}{1+n} \qquad R = \frac{k-k'}{k+k'} = \frac{1-n}{1+n}$$

Exercise 4

A uniform string of mass m and length l hangs in a field of gravity g. How do waves propagate on the string?

The tension in the string is:

$$\tau = mgz/\ell,$$

The speed of the waves is:

$$v(z) = \sqrt{\tau/\lambda} = \sqrt{gz}$$

If the wavelength is much less than l, the waves propagate diabatically:

$$t(\ell \to z) = \int_z^\ell \frac{dz}{\sqrt{gz}} = 2[\sqrt{\ell/g} - \sqrt{z/g}]$$

Exercise 5

A cylinder of mass M and radius R rests on a horizontal plane. The cylinder axis is connected to a fixed point by means of a spring of elastic constant k, initially at rest.
The cylinder is moved from the equilibrium position and released.
Determine the equation of motion in the hypothesis that the motion of the cylinder is pure rolling.

The spring force is given by:

$$F_{el} = -kx$$

With respect to the center of rotation Q has a moment equal to:

$$\tau_q = \tau_{el} = kxR$$

The moment of inertia of the cylinder is:

$$I_Q = 3\,MR^2/2 \;.$$

The center of mass has a relationship between angular and linear accelerations like this:

$$\alpha = -\frac{a}{R} = -\frac{1}{R}\frac{\mathrm{d}^2 x}{\mathrm{d}t^2} \;.$$

The equation of motion is:

$$\frac{\mathrm{d}^2 x}{\mathrm{d}t^2} + \frac{2k}{3M}x = 0 \;.$$

And it is a simple harmonic motion of period:

$$T = 2\pi\sqrt{3M/2k} \;.$$

Exercise 6

A plate of area A immersed in a fluid is suspended from a spring of spring constant k.
Assume that the frictional force is proportional to the speed of the plate and to 2A.
Write the equation of motion of the plate.

Vertical forces act on the plate: the weight force, the Archimedean thrust, the elastic force and the friction force, shown below:

$$P = mg,$$
$$S = -\rho_l V_l g;$$
$$F_e = -kx';$$
$$F_a = -2\mu A(dx'/dt).$$

Where mu is the coefficient of viscous friction.
The equation of motion is:

$$\frac{d^2 x'}{dt^2} + \frac{2\mu A}{m} \frac{dx'}{dt} + \frac{k}{m} x' = g - \frac{\rho_l V_l}{m} g .$$

Whose solutions are:

$$x(t) = x_0 e^{-\gamma t} \sin(\omega_s t + \phi) , \qquad \omega_s = \sqrt{\omega_0^2 - \gamma^2} .$$

$$\gamma = \mu A/m,$$

$$\omega_0^2 = k/m$$

It is a period-damped harmonic motion:

$$T = \frac{2\pi}{\omega_s} = \frac{2\pi}{\sqrt{k/m - \mu^2 A^2/m^2}} .$$

Exercise 7

A spring AB of spring constant k is connected at B to a body of mass m
sliding along a frictionless horizontal plane.
The end point A of the spring is bound by the clockwise law:

$$X_A = X_0 \sin(\omega t)$$

Determine the force applied to the body mass m and its hourly law.

231

The force acts on the body:

$$F_B = k(X_A - X_B) .$$

The equation of motion is:

$$m \, \frac{d^2 X_B}{dt^2} = F_B = k(X_A - X_B) .$$

Place:

$$X_A = X_0 \sin(\omega t), \; \omega_0^2 = k/m$$
$$\rho_0 = kX_0/m$$

You get:

$$\frac{d^2 X_B}{dt^2} + \omega_0^2 X_b = \rho_0 \sin(\omega t) .$$

Whose solution is:

$$X_B = D \sin(\omega_0 t + \phi) + \frac{\rho_0}{\omega_0^2 - \omega^2} \sin(\omega t) .$$

It is a forced harmonic motion.

9

THERMODYNAMICS AND HEAT TRANSMISSION

Historical evolution

The study of thermodynamics is one of those cases in modern physics which contradict the logical principle of cause and effect between theory and practice.

In fact, first there were the practical implications, such as the steam engine and the application of then unknown thermodynamic concepts and only later was a complete theoretical explanation given.

Among other things, the extent of the changes related to the application of the steam engine was of such considerable dimensions in social and economic terms that the term Industrial Revolution was coined to the period comprising the first three decades of the nineteenth century.

The first scientist who systematically studied thermodynamics was Carnot, to whom we owe the demonstration in 1824 of Carnot's theorem, the Carnot cycle and Carnot's ideal machine.

The same introduced the concepts of thermodynamic efficiency and work as heat exchange.

Subsequently Joule demonstrated, in 1850, the equality between the forms of energy deriving from heat and mechanical work, paving the way for the first law of thermodynamics.

In a parallel way, Kelvin and Clausius began to probe the differences between heat and work, coming to enunciate, separately, the second law of thermodynamics right at the turn of the mid-nineteenth century.

We owe the concepts of thermodynamic temperature to Kelvin and to Clausius the distinction between reversible and irreversible processes, with the introduction of the entropy state function.

Finally, in the second half of the 19th century, Maxwell abstracted general relations for thermodynamics and Gibbs, in 1876, introduced the description of state variables.

Definitions

The basic concepts of thermodynamics are those of volume, pressure and temperature derived from other disciplines (such as mechanics and fluid theory).

In thermodynamics, the centigrade scale does not apply, which makes the freezing point of water at atmospheric pressure correspond to zero degrees and the boiling point of water to one hundred degrees, again at atmospheric pressure.

To measure the temperature, the Kelvin scale is used and the temperature expressed in this way is called absolute.

The Kelvin scale makes the value of the absolute minimum temperature correspond to zero (which is equal to -273.15 °C).

This temperature, called absolute zero, is the minimum possible and we will clearly see it by enunciating the laws of thermodynamics.

To be precise, the reference point of the Kelvin scale is not absolute zero, but the triple point temperature of water at atmospheric pressure which is equal to 273.16 K.

Thermodynamics is based on the mathematical formalism of partial derivatives and this can be seen immediately by defining the thermodynamic coefficients:

Isochoric pressure coefficient:

$$\beta_V = \frac{1}{p}\left(\frac{\partial p}{\partial T}\right)_V$$

Isothermal compressibility:

$$\kappa_T = -\frac{1}{V}\left(\frac{\partial V}{\partial p}\right)_T$$

Volume isobar coefficient:

$$\gamma_p = \frac{1}{V}\left(\frac{\partial V}{\partial T}\right)_p$$

Adiabatic Compressibility:

$$\kappa_S = -\frac{1}{V}\left(\frac{\partial V}{\partial p}\right)_S$$

By definition, an isochoric transformation keeps the volume constant, an isotherm the temperature, an isobaric the pressure, an adiabatic the entropy (which we will define later). An adiabatic transformation does not exchange heat with the external system.

For an ideal gas, the following relationship between pressure, volume and temperature holds (ideal gas law):

$$pV = nRT$$

Where n is the number of moles and R is the universal gas constant.
In this case, the above coefficients are simply the following:

$$\gamma_p = \frac{1}{T}$$

$$\kappa_T = \frac{1}{p}$$

$$\beta_V = -\frac{1}{V}$$

The global view of thermodynamics generalizes the concept of specific heat, introduced at the dawn of heat studies to explain the different thermal capacities of materials, i.e. the different temperature changes following the same heat source.

Specific heat at a generic constant X:

$$C_X = T\left(\frac{\partial S}{\partial T}\right)_X$$

Specific heat at constant pressure:

$$C_p = \left(\frac{\partial H}{\partial T}\right)_p$$

Specific heat at constant volume:

$$C_V = \left(\frac{\partial U}{\partial T} \right)_V$$

H indicates the enthalpy and U the internal energy; both quantities will be described shortly.

We define specific molar heats as the specific heats per mole of substance. For an ideal gas, the following relation holds for the specific molar heats:

$$C_{mp} - C_{mV} = R$$

In general, one can obtain:

$$C_p - C_V = T \left(\frac{\partial p}{\partial T} \right)_V \cdot \left(\frac{\partial V}{\partial T} \right)_p = -T \left(\frac{\partial V}{\partial T} \right)_p^2 \cdot \left(\frac{\partial p}{\partial V} \right)_T \geq 0$$

Since

$$\left(\frac{\partial p}{\partial V} \right)_T$$

is less than zero, then we necessarily have that

$$C_p \geq C_V$$

The equality between the two specific heats occurs only if the expansion coefficient is zero or if the temperature is zero.

The first law of thermodynamics

The zeroth law of thermodynamics states that heat flows from systems of higher temperature to systems of lower temperature.

The first law of thermodynamics is an extension of the law of conservation of energy.
Having defined Q the heat flow, W the work done and U the internal energy of a system, the first law is expressed in this way, for a closed system:

$$Q = \Delta U + W$$

In differential form, we have to make a distinction between exact and inexact differentials.

The functions for which an exact differential can be expressed are called state functions and characterize the thermodynamic state of a system.

There are four state functions of a thermodynamic system which describe and characterize it completely: internal energy, enthalpy, free energy and Gibbs free enthalpy.

Apart from the internal energy, which by definition is U, the other quantities are:

$$H = U + pV$$
$$F = U - TS$$
$$G = H - TS$$

In differential form we have:

$$dU = TdS - pdV$$
$$dH = TdS + Vdp$$
$$dF = -SdT - pdV$$
$$dG = -SdT + VdP$$

We rewrite the first law in differential form:

$$\delta Q = dU + \delta W$$

For a quasi-static and reversible process the following expressions hold:

$$\delta W = pdV \Rightarrow \delta Q = dU + pdV$$

For an open system, the first law of thermodynamics is very reminiscent of the conservation of mechanical energy, provided that we remember the contributions linked to heat and chemical energy (which can be expressed with the enthalpy state function):

$$Q = \Delta H + W + \Delta E_{cin} + \Delta E_{pot}$$

The second law of thermodynamics

The real leap in thermodynamics, however, was given by the second law.
The origin of this principle derives from empirical considerations: there is
no perfect equivalence between the concepts of heat and work.
Put another way, in real cases it is possible to convert some or all of the
work into heat, but it is not possible to do the other way around.
Hence the definition of irreversible transformations since it is not possible
to return to the same starting thermodynamic state.
Heat is the most disordered and degenerate form of energy, causing a loss
of information about the thermodynamic state.
To explain these experimental evidences, a state variable was introduced,
called entropy, and denoted by S.
For a reversible process we have:

$$dS = \frac{\delta Q_{rev}}{T} \Rightarrow S_2 - S_1 = \int_1^2 \frac{\delta Q_{rev}}{T} \Rightarrow \oint \frac{\delta Q_{rev}}{T} = 0$$

For an irreversible process the following holds instead:

$$dS > \frac{\delta Q_{irr}}{T} \Rightarrow S_2 - S_1 > \int_1^2 \frac{\delta Q_{irr}}{T} \Rightarrow \oint \frac{\delta Q_{irr}}{T} < 0$$

Between a final and an initial thermodynamic state, entropy cannot
decrease.
In real cases, entropy always increases, as soon as heat quantities and
flows are involved.

The consequences of this principle are numerous.
The first is that the existence of perpetual motions or perpetual
thermodynamic cycles is totally excluded, ie that certain processes and
certain transformations can be repeated indefinitely without any change in
the thermodynamic conditions of the system.
There is always a price to pay, precisely given by the increase in entropy,
when thermodynamic work is done.
Other practical consequences are the limitation of the yield and
thermodynamic efficiency which, by definition, can never be unitary.

The third law of thermodynamics

The third law of thermodynamics states that entropy vanishes only at the absolute zero limit:

$$\lim_{T \to 0} \left(\frac{\partial S}{\partial X} \right)_T = 0$$

From this principle it follows that the heat capacity and specific heats go to zero when the temperature of absolute zero is reached.
The immediate consequence is that it is not possible to reach absolute zero, even with infinite successive steps of cryogenic cooling.
This value remains an unattainable and not physically reproducible limit.

Maxwell and Helmholtz relations

From the definitions of the state functions in differential form, the Maxwell relations can be obtained:

$$\left(\frac{\partial T}{\partial V} \right)_S = -\left(\frac{\partial p}{\partial S} \right)_V$$

$$\left(\frac{\partial T}{\partial p} \right)_S = \left(\frac{\partial V}{\partial S} \right)_p$$

$$\left(\frac{\partial p}{\partial T} \right)_V = \left(\frac{\partial S}{\partial V} \right)_T$$

$$\left(\frac{\partial V}{\partial T} \right)_p = -\left(\frac{\partial S}{\partial p} \right)_T$$

Which can also be expressed in this way with the specific heats:

$$TdS = C_V \cdot dT + T\left(\frac{\partial p}{\partial T}\right)_V dV = C_p dT - T\left(\frac{\partial V}{\partial T}\right)_p dp$$

For an ideal gas, the simplified relation for molar entropies holds:

$$S_m = C_V \ln\left(\frac{T}{T_0}\right) + R\ln\left(\frac{V}{V_0}\right) + S_{m0} = C_p \ln\left(\frac{T}{T_0}\right) - R\ln\left(\frac{V}{V_0}\right) + S'_{m0}$$

The following Helmholtz equations generalize the Maxwell relations:

$$\left(\frac{\partial U}{\partial V}\right)_T = T\left(\frac{\partial p}{\partial T}\right)_V - p$$

$$\left(\frac{\partial H}{\partial p}\right)_T = V - T\left(\frac{\partial V}{\partial T}\right)_p$$

Thermodynamic processes

To describe thermodynamic processes it is necessary to introduce efficiency as the ratio between work performed and heat input.
In the case of cooling processes, the corresponding factor is also useful, which is the ratio of the cooling heat flux to the work done.
Defined K as a generic constant, for a reversible adiabatic process the following holds:

$$\gamma = \frac{C_P}{C_V} \Rightarrow pV^\gamma = TV^{\gamma-1} = T^\gamma p^{1-\gamma} = K$$

For a reversible isobaric process, the following holds:

$$\Delta H = Q_{rev}$$

The Carnot cycle or process consists of a reversible cycle made up of four transformations given by: an isothermal expansion at a given temperature, an adiabatic expansion which brings the system to a lower temperature, an isothermal compression at this lower temperature and finally a compression adiabatic which brings the system back to the initial conditions.

During isothermal expansion the system absorbs a certain amount of heat, during isothermal compression it releases another.

Setting the subscript 1 to the first quantity described and the subscript 2 to the second, the efficiency of the Carnot cycle is given by:

$$\eta = 1 - \frac{|Q_2|}{|Q_1|} = 1 - \frac{T_2}{T_1} = \eta_C$$

The Carnot efficiency is the maximum possible for any thermodynamic process.

If the cycle is performed in the opposite direction, the cooling factor can be defined as:

$$\xi = \frac{T_2}{T_1 - T_2}$$

In thermodynamics there are many ideal cycles and processes that give rise to numerous practical applications of general interest, such as for example refrigeration systems and internal combustion engines.

Those cycles which transform usable heat energy into mechanical energy are said to be direct, while if the opposite occurs, the cycles are said to be inverse.

The latter can supply heat at temperatures above ambient temperature (heat pumps) or they can extract heat at temperatures below ambient temperature (refrigerators).

The Stirling cycle is composed of two isothermal and two isochoric transformations, each of which can be divided into expansion and compression.

This cycle has an efficiency equal to that of the Carnot cycle.

The Otto cycle is composed of two adiabatic transformations and two isochores, each of which can be divided into expansion and compression.

This cycle is the idealization of the functioning of an internal combustion engine fueled by petrol and has a lower efficiency than that of Carnot.

A compression ratio can be defined for this cycle.

The Joule cycle is the idealization of what is achieved in gas turbines and consists of two adiabatic transformations and two isobars, each of which can be divided into expansion and compression.

The Diesel cycle is instead the idealization of what happens in compression ignition engines, such as those fueled by diesel.
It is composed of two adiabatic transformations, one of compression and one of expansion, of an expansion isobar and a compression isochore.

Thermodynamic cycles can be combined and can involve not only gases but also fluids, as in the case of the Rankine cycle.

The maximum work that can be obtained from a reversible process at a given temperature and pressure is expressed as follows for closed and open systems:

Closed systems:

$$W_{max} = (U_1 - U_2) - T_0 (S_1 - S_2) + p_0 (V_1 - V_2)$$

Open systems:

$$W_{max} = (H_1 - H_2) - T_0 (S_1 - S_2) - \Delta E_{cin} - \Delta E_{pot}$$

Multi-phase systems and thermodynamic potentials

A particular thermodynamic process is that given by phase transitions which are, by definition, isotherms and isobars for which dG=0.
The following relations are therefore valid for the molar quantities (we denote the three phases with the letters alpha, beta and gamma):

$$G_m^\alpha = G_m^\beta$$

$$\Delta S_m = \frac{r_{\beta\alpha}}{T_0}$$

$$r_{\beta\alpha} = r_{\alpha\beta} = r_{\gamma\alpha} - r_{\gamma\beta}$$

where T_0 is the transition temperature and $r_{\beta\alpha}$ is the heat of transition between two phases.
From these relations, the expression for molar entropy is easily derived:

$$S_m = \left(\frac{\partial G_m}{\partial T} \right)_p$$

For a two-phase system, the Clapeyron equation holds:

$$\frac{dp}{dT} = \frac{r_{\beta\alpha}}{\left(V_m^{\alpha} - V_m^{\beta} \right) T}$$

For an ideal gas at a distance from the critical point, a pressure relation holds:

$$p = p_0 e^{-\frac{r_{\beta\alpha}}{RT}}$$

There are also more complex phase transitions, for example of the second order for the phenomena of reorganization of matter or of the third order, as in the case of the phase change of iron when it passes from ferromagnetic to paramagnetic behavior.

When the number of particles of a thermodynamic system changes within a process, then this number becomes a new state quantity, which can be described by a state function.
Since these processes take place at constant pressure and temperature, it is the Gibbs free enthalpy that incorporates this change:

$$dG = -SdT + Vdp + \sum_i \mu_i dn_i$$

Having defined the thermodynamic potential as follows:

$$\mu = \left(\frac{\partial G}{\partial n_i} \right)_{p,T,n_j}$$

In a multiphase system, the thermodynamic potentials are not independent of each other.
The Gibbs-Duhmen relation establishes the constraints that thermodynamic potentials must satisfy:

$$\sum_i \frac{n_i}{n} d\mu_i = 0$$

A system with n components and p phases has a number of free parameters given by

$$f = n + 2 - p$$

A mixture of n different components has an internal energy and an enthalpy given by the weighted sum of these quantities for each single component, while the entropy is increased by a quantity due to the mixture itself. In formulas we have:

$$U_{misc} = \sum_i n_i U_i^0$$

$$H_{misc} = \sum_i n_i H_i^0$$

$$S_{misc} = n \sum_i \frac{n_i}{n} S_i^0 + \Delta S_{mix}$$

For an ideal gas the extra contribution on the entropy is:

$$\Delta S_{mix} = -nR \sum_i \frac{n_i}{n} \ln\left(\frac{n_i}{n}\right)$$

The thermodynamic potentials of a mixture are lower than the equivalents of the individual compounds:

$$\mu_i = \mu_i^0 + RT \ln\left(\frac{n_i}{n}\right) < \mu_i^0$$

A solution of one compound in another causes a boiling point elevation and a freezing point depression.
If either compound is not substantially present, linear relationships hold with respect to mole fractions.

$$\Delta T_{eb} = \frac{RT_{eb}^2}{r_{\beta\alpha}} \frac{n_2}{n}$$

$$\Delta T_{cong} = -\frac{RT_{cong}^2}{r_{\gamma\beta}} \frac{n_2}{n}$$

Even in thermodynamics it is possible to describe the system around equilibrium. In this situation, the possible oscillations of the thermodynamic state are those in which the following relations hold:

$$(dS)_{U,V} \geq 0 \leftrightarrow (dU)_{S,V} \leq 0 \leftrightarrow (dH)_{S,P} \leq 0 \leftrightarrow (dF)_{T,V} \leq 0 \leftrightarrow (dG)_{T,P} \leq 0$$

In the equilibrium condition the equality of the thermodynamic potentials always holds:

$$\mu_i^\alpha = \mu_i^\beta = \mu_i^\gamma$$

Statistical view

We anticipate a statistical consideration about entropy by referring to the next chapter for a complete understanding.
Said P the total number of possibilities of distributing N particles in n different energy levels, each of which can be degenerated g times, we have that:

$$P = N! \prod_i \frac{g_i^{n_i}}{n_i!}$$

This quantity is called the thermodynamic probability.
The most probable distribution is the one that maximizes the thermodynamic probability and is given by the equilibrium state.
We will see in the next chapter how this distribution is that of Maxwell-Boltzmann.
The statistical vision of entropy indicates how this state function "measures" the disorder of the thermodynamic system.
The higher the P value, the more disorder there will be and therefore the more entropy will be associated with that thermodynamic state.

In formulas, one has.

$$S = k\ln(P)$$

With k a suitable constant.
Defined Z as a normalization constant of the distribution, in the equilibrium condition it follows that:

$$S = \frac{U}{T} + kN\ln\left(\frac{Z}{N}\right) + kN$$

For an ideal gas at equilibrium, the following result holds:

$$S = \frac{5}{2}kN + kN\ln\left(\frac{V(2\pi mkT)^{\frac{3}{2}}}{Nh^3}\right)$$

Heat transfer

The heat transfer can take place through three different physical processes. We speak of convection when the transmission of heat takes place with the exchange of matter, for example between a surface and a moving fluid having different temperatures.
Conduction, on the other hand, takes place without the exchange of matter when there is a temperature gradient in a stationary medium, whether solid or fluid.
Finally, radiation is caused by the emission of energy in the form of electromagnetic waves and can occur both in contact and through interposed means, i.e. at a distance, as in the case of heat from the Sun.
In some cases, the heat can be transmitted simultaneously by radiation and convection and then we speak of adduction.

As far as convection is concerned, the laws of heat transmission are those of fluid dynamics, in particular everything including the Navier-Stokes equations.

For the conduction the Fourier equation is valid where, defined q the thermal current density, we have:

$$\frac{\partial q}{\partial t} = -\sigma_{\mu\nu} \nabla T$$

$\sigma_{\mu\nu}$ is the thermal conductivity tensor which reduces to the thermal conductivity constant if the medium is homogeneous and isotropic.
Another way of expressing Fourier's equation, particularly convenient for solids, is:

$$\nabla^2 T = \frac{1}{D_Q}\frac{\partial T}{\partial t}$$

Where is indicated the thermal diffusivity given by the ratio between thermal conductivity and the product between density and specific heat:

$$D_Q = \frac{\sigma}{\rho c_p}$$

To describe the physical laws on radiation it is instead necessary to resort to electromagnetism.
There are empirical results regarding irradiation.
The first is that of Kirchhoff according to which the sum of the coefficients of transmission, absorption and reflection is equal to one, having to submit to the conservation of energy.
From this principle, we can define a black body as a body which absorbs all the incident radiation, without reflecting or transmitting anything.
For a black body, the Stefan-Boltzmann law holds, according to which the global emittance of a black body is given by

$$J = \sigma T^4$$

where σ is the Stefan-Boltzmann constant.
To understand the importance of thermodynamic radiation, we can state that theoretically one of the first denials of all of modern physics as presented up to now has taken place precisely by studying black bodies, while practically the theory of radiation explains the flow of heat which from the Sun it propagates towards la Terra, guaranteeing the existence of all life on our planet.

High school exercises

Exercise 1

Find the volume change of an aluminum sphere of radius 10 cm when heated from 0 to 100°C.

Knowing that the linear coefficient of thermal expansion of aluminum is:

$$\alpha_{Al} = 23 \cdot 10^{-6} \, {}^{\circ}C^{-1}.$$

The volumetric variation is given by:

$$\Delta V = 3 V_0 \alpha \Delta T$$

And then:

$$\Delta V = 3 \times \frac{4}{3} \pi \times 1000 \, cm^3 \times 23 \cdot 10^{-6} \, {}^{\circ}C^{-1} \times 100 \, {}^{\circ}C = 29 \, cm^3$$

Exercise 2

Calculate the amount of heat required to bring 1 kg of water from the temperature of 0°C to that of 25°C.

The relationship is given by:

$$Q = mc_s \, \Delta T$$

Where the specific heat of water is:

$$c_{acqua} = 4180 \, J/kg \, K,$$

And then:

$$Q = 1 \, kg \cdot 4180 \, \frac{J}{kg \, K} \cdot 25 \, K = 104500 \, J$$

In calories:

$$Q = 1000\,g \cdot 1,01\,\frac{cal}{g\,K} \cdot 25 = 25250\,cal = 25,25\,kcal$$

Exercise 3

A certain substance has a molar mass of 50 g/mol. By supplying 314 J to 30 grams of this substance, its temperature increases from 25°C to 45°C. Find the specific heat of the substance, the number of moles and the molar specific heat.

The heat absorption follows this law:

$$Q = c_s m \Delta T$$

And then we can find the specific heat:

$$c_s = \frac{314}{20 \times 0.030} = 523\,\frac{J}{kg\,°C}$$

The number of moles is given by the ratio between the amount of substance and the molar mass:

$$n = \frac{m}{M} = \frac{30}{50} = 0.600\,moli$$

Molar specific heat is the product of specific heat and molar mass:

$$c_s^{molare} = 523\frac{J}{kg\,°C} \times 0.05\,\frac{kg}{mole} = 26.2\,\frac{J}{mole\,°C}$$

Exercise 4

A metal container has a mass of 3.6 kg and contains 14 kg of water.

249

A piece of the same metal with a mass of 1.8 kg is placed in water at a temperature of 180°C.
The container and the water have a temperature of 16°C at the beginning and 18°C at the end.
Find the specific heat of the metal.

By the principle of conservation of energy:

$$Q_{cont} + Q_{acqua} + Q_{metallo} = 0$$

That is, all the heat released by the metal is absorbed by the water and the container.
Substituting, we get:

$$3.6 \times c_s \times 2 + 14 \times 4186 \times 2 - 1.8 \times c_s \times 162 = 0$$

From which:

$$c_s = 412 \frac{J}{kg\,K}$$

Exercise 5

A sample of gas expands from 1 to 4 cubic meters along path B of the pV diagram in the figure:

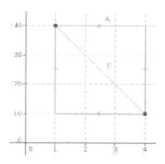

The gas is recompressed to 1 cubic meter following either path A or C.

250

Calculate the work done by the gas during the complete cycle in both cases.

On path B the work done by the gas is:

$$W = \frac{(40 + 10) \times 3}{2} = 75 \, J$$

If the cycle closes following path A, the gas pressure first increases for the same volume and therefore in this section the work is zero.
Subsequently the volume decreases at the same pressure and therefore:

$$W = -p\Delta V = -40 \times 3 = -120 \, J$$

The work in the BA cycle will be:

$$W_{BA} = 75 - 120 = -45 \, J$$

If, on the other hand, the cycle closes following path C, in the first section there is a compression of the volume at the same pressure and therefore:

$$W = -10 \times 3 = -30 \, J$$

While in the second section the volume is constant and therefore the work is zero.
The work in the BC cycle will be:

$$W_{BC} = 75 - 35 = 45 \, J.$$

Exercise 6

A gas follows this cycle:

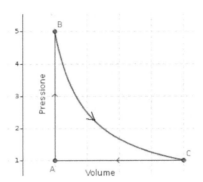

Determine the total heat supplied to the system during the AC transformation assuming that the heat supplied during the AB transformation is 20 J, that the BC transformation is adiabatic, and that the total work done during the cycle is 150 J.

The transformation AB produces no work as it is isochorous.
The BC transformation does not exchange heat.
By the first law of thermodynamics it must be that all work is transformed into heat.
Or:

$$W_{tot} = 15.0\ J,$$
$$Q_{tot} = 15.0\ J$$

But:

$$Q_{AB} = 20.0\ J$$

Therefore:

$$Q_{CA} = -5\ J$$

Exercise 7

The heat flux per unit time through the surface of the Earth in continental areas is:

$$54.0 \, mW/m^2,$$

And the thermal conductivity of the surface is:

$$2.50 \, W/ \, (m \cdot K).$$

Assuming that the surface temperature is 10 °C, what is the temperature at a depth of 35 km if there are no internal heat generation mechanisms?

The heat flux per unit time is given by:

$$H = \frac{Q}{t} = kA\frac{T_1 - T_2}{L}$$

Where L is the thickness of the material, A and k the thermal conductivity. Therefore:

$$T_1 - T_2 = \frac{54.0 \cdot 10^{-3} \frac{W}{m^2} \times 3.5 \cdot 10^4 \, m}{2.50 \frac{W}{m \cdot K}} = 756$$

The temperature at 35 km will be 766°C.

Exercise 8

A sphere with a radius of 0.5 meters, an emissivity of 0.85 and a temperature of 27°C is placed in an environment with a temperature of 77°C.
Find the radiant power that the sphere emits, absorbs, and the net power exchanged.

The power emitted by the sphere by radiation is given by the Stefan-Boltzmann law:

$$P_r = \sigma \varepsilon A T^4$$

Therefore:

$$P_r = 5.6703 \cdot 10^{-8} \, \frac{W}{m^2 \, K^4} \times 0.850 \times 3.14 \, m^2 \times 300^4 \, K^4 = 1226 \, W$$

Since the surface area of the sphere is equal to:

$$4\pi r^2 = 3.14 \, m^2$$

The power absorbed from the environment is:

$$\bar{P}_a = \sigma \varepsilon A T_{amb}^4.$$

Therefore:

$$P_a = 5.6703 \cdot 10^{-8} \, \frac{W}{m^2 \, K^4} \times 0.850 \times 3.14 \, m^2 \times 350 \, K = 2271 \, W$$

The net power exchanged:

$$P_n = P_a - P_r = 2271 - 1226 = 1045 \, W$$

Exercise 9

An object with a surface area of 200 square centimeters and a constant temperature of 70°C is immersed in a fluid with a constant temperature of 20°C.
Calculate the external thermal conductivity coefficient assuming that the object is able to disperse a quantity of heat equal to 1'040'000 Kcal in the fluid in 12 hours.

The amount of heat lost is given by:

$$Q = \lambda_e \, (T_2 - T_1) \, St,$$

Substituting, we get:

$$1040 \cdot 10^3 \, kcal = \lambda_e \, \frac{kcal}{h \, m \, °C} \cdot 50 \, °C \cdot 200 \cdot 10^{-4} \, m^2 \cdot 12 \, h$$

And then:

$$\lambda_e = \frac{1040 \cdot 10^3 \, kcal}{12 \, h \, m \, ^\circ C} = 86667, 7 \, \frac{kcal}{h \, m \, ^\circ C}$$

Exercise 10

A cylinder with a volume of 100 cubic centimeters contains an ideal gas at a pressure of 10,000,000 N per square metre.
Assuming that the temperature remains constant during the process, calculate how many balloons can be filled if each balloon has a volume of 15 cubic centimeters and the filling pressure is 180,000 N per square metre.

From Boyle's law at constant temperature:

$$V_1 = \frac{p_0}{p_1} \cdot V_0 = \frac{10^7 \, \frac{N}{m^2}}{1, 8 \cdot 10^5 \, \frac{N}{m^2}} \cdot 10^{-4} m^3 = 5, 56 \cdot 10^{-3} m^3$$

And so the number of balloons is:

$$n_{palloncini} = \frac{5560 \, cm^3}{15 \, cm^3} = 370$$

Exercise 11

10 moles of ideal gas at the initial pressure of 5 atmospheres contained in a volume of 16.42 liters are expanded to triple the volume.
Assuming that the pressure remains constant, calculate the final temperature of the gas.

In the international system we have:

$$p = 5 \, atm = 5 \cdot \left(1, 013 \cdot 10^5\right) \frac{N}{m^2}$$

From the ideal gas equation of state:

$$5 \cdot \left(1,013 \cdot 10^5\right) \frac{N}{m^2} \cdot 3 \cdot \left(16,42 \cdot 10^{-3}\right) m^3 = 10 \cdot 8,31 \frac{N}{K\,m} \cdot T$$

Solving for T:

$$T = \frac{24950\,Nm}{10 \cdot 8,31\ \frac{J}{K\,mole}} = 300\,K$$

Exercise 12

An ideal gas undergoes a reversible isothermal expansion at 132°C. The entropy of the gas increases to 46 J/K.
Find the amount of heat absorbed.

In a reversible isothermal transformation we have:

$$\Delta S = \frac{Q}{T}$$

And then:

$$Q = \Delta S \cdot T = 46.0\,\frac{J}{K} \times 405\,K = 18680\,J$$

Exercise 13

Given two heat sources at 400 K and 100 K, we transfer 260 J from the first to the second source by conduction.
Calculate the change in entropy.

The variation is simply given by:

$$\Delta S = \frac{Q_f}{T_f} - \frac{Q_i}{T_i} = \frac{260}{100}\left(1 - \frac{1}{4}\right) = 1.95 \; J$$

Exercise 14

Mix 200 grams of aluminum at 100°C with 50 grams of water at 20°C. Calculate the equilibrium temperature, the entropy change of aluminum and water and the entropy change of the system.

From the law of thermodynamics all the heat released by the aluminum is absorbed by the water:

$$m_{Al}c_s^{Al}\Delta T + m_{acqua}c_s^{acqua}\Delta T = 0$$

At equilibrium we have:

$$0.200 \; kg \times 900 \frac{J}{kg \cdot K} \times (100 - T_{eql}) = 0.050 \; kg \times 4186 \frac{J}{kg \cdot K} \times (T_{eql} - 20)$$

For which:

$$(209.3 + 180) \, T_{eql} = 18000 + 4186$$

$$T_{eql} = \frac{22186}{389.3} = 57.0 \, °C$$

The entropy changes of aluminum, water and the system are:

$$\Delta S_1 = mc_{Al} \ln\left(\frac{330}{373}\right) = -22.1 \frac{J}{K}$$

$$\Delta S_2 = mc_{acqua} \ln\left(\frac{330}{293}\right) = 24.9 \frac{J}{K}$$

$$\Delta S = \Delta S_1 + \Delta S_2 = -22.1 + 24.9 = 2.8 \frac{J}{K}$$

Exercise 15

Calculate the efficiency of a machine that consumes 380 tons of coal every hour to produce useful work at a power of 750 MW.
Consider the heat of combustion of coal equal to 28 MJ per Kg.

The amount of heat consumed each hour is:

$$Q = 2.8 \cdot 10^7 \, \frac{J}{kg} \times 3.80 \cdot 10^5 \, kg = 1.06 \cdot 10^{13} \, J$$

The useful work is given by:

$$W = P\Delta t = 7.50 \cdot 10^8 \, \frac{J}{s} \times 3600 \, s = 2.7 \cdot 10^{12} \, J$$

The yield is therefore equal to:

$$\eta = \frac{W}{Q} = \frac{2.7 \cdot 10^{12} \, J}{1.06 \cdot 10^{13} \, J} = 0.25 = 25\%$$

Exercise 16

How much energy does it take to raise the temperature of a mass of 10 kg of water from 80°C to 130°C?

The energy is given by the sum of three terms.
The first is needed to bring the water from 80°C to 100°C (the boiling temperature).
The second term is the latent boiling energy, i.e. the energy required to make the water boil completely.
The third term is the one needed to bring the water from 100°C to 130°C.
The first term is given by:

$$\Delta Q_1 = c_s \, m \, \Delta T$$

$$\Delta Q_1 = 4186 \frac{J}{kg \, K} \, 10 kg \, 20 K = 837, 2 \, kJ$$

The second instead:

$$\Delta Q_2 = Q_{lat-eb} \cdot m = 2272 \frac{kJ}{kg} \cdot 10 \, kg = 22720 \, kJ$$

And the third:

$$\Delta Q_3 = c_s \, m \, \Delta T$$

$$\Delta Q_3 = 4186 \frac{J}{kg \, K} \, 10 kg \, 30 K = 1255, 8 \, kJ$$

The total amount of energy is:

$$\Delta Q_{tot} = \Delta Q_1 + \Delta Q_2 + \Delta Q_3 = 24813 \, kJ$$

We note how the prominent contribution is that of the latent boiling energy.

Exercise 17

Given a gas at 380 K it undergoes an isobaric transformation going from 10 cubic centimeters of volume to double, what temperature has it reached?

The ideal gas law applied to the initial and final state leads to:

$$\begin{cases} PV_f = NKT_f \\ PV_i = NKT_i \end{cases}$$

Therefore:

$$\frac{V_f}{V_i} = \frac{T_f}{T_i}$$

Or:

$$T_f = \frac{20\,cm^3 \cdot 380\,K}{10\,cm^3} = 760\,K$$

Exercise 18

A gas completes a thermodynamic cycle formed by two isobars and two isochores as in the figure:

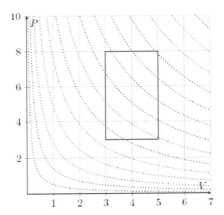

The cycle begins with an isobaric transformation at 8 atm which brings the volume from 3 to 5 cubic metres, then an isochoric cooling followed by an isobaric compression at 3 atm which brings the volume back from 5 to 3 cubic metres.
Calculate the work done in each transformation and the total cycle work.

The work done in the two isochores is zero.
In isobar expansion the work is:

$$L = P \cdot \Delta V = 8\,atm \cdot 2\,m^3 = 1600000\,J$$

In isobar compression it is:

$$L = P \cdot \Delta V = 3\,atm \cdot (-2\,m^3) = -600000\,J$$

The total work is:

$$L = 1000000\,J$$

Exercise 19

A thermodynamic cycle absorbs heat at high temperatures, releases it at low temperatures and releases work.
Knowing that the heat absorbed is 5'000 J and that transferred 3'500 J, calculate the work and the efficiency of the cycle.

In formulas we have:

$$\delta Q_{ass} = \delta Q_{ced} + \delta L$$

$$\eta = \frac{\delta L}{\delta Q_{ass}}$$

I our case:

$$\delta L = \delta Q_{ass} - \delta Q_{ced} = 1500\,J$$

$$\eta = \frac{\delta L}{\delta Q_{ass}} = \frac{1500\,J}{5000\,J} = 0,3 = 30\%$$

Exercise 20

A Carnot cycle absorbs 1,000 J at 1,000 K and gives off heat at 400 K. What is its yield? How much work is produced?

In a Carnot cycle, the efficiency depends only on the temperatures within which it works:

$$\eta = 1 - \frac{T_{bassa}}{T_{alta}} = 1 - \frac{4}{10} = \frac{6}{10} = 0,6 = 60\%$$

So the work will be:

$$\delta L = \eta \delta Q_{ass} = 0,6 \cdot 1000 \, J = 600 \, J$$

Exercise 21

What is the minimum amount of work that must be used in a Carnot cycle to subtract 180 J from a gas at a temperature of -3°C in an environment at 27°C?

After transforming the temperatures into Kelvin, the efficiency of the cycle is:

$$\eta_c = 1 - \frac{T_b}{T_a} = 1 - \frac{270}{300} = 0,1$$

You will have:

$$\begin{cases} \delta L = \delta Q_{Ta} \cdot \eta_c \\ \delta Q_{Tb} = \delta Q_{Ta} - \delta L \end{cases}$$

Doing the accounts:

$$\delta Q_{Tb} = \delta L \cdot \left(\frac{1}{\eta_c} - 1 \right)$$

$$\delta L = \delta Q_{Tb} \cdot \frac{\eta_c}{1 - \eta_c} = 10 \, J \cdot \frac{0,1}{0,9} = 20 \, J$$

Exercise 22

A mass of 560 grams of nitrogen, whose molecular weight is 28 grams per mole, is found at 270 K and is contained in a cylinder with a section of 1,000 square centimeters and a height of 1 meter.
What pressure is the gas at?
What if its temperature rises by 30°C?

The number of nitrogen molecules is given by:

$$N = \frac{m}{PM} \cdot N_A = \frac{560\,g}{28\,\frac{g}{mole}} \cdot 6,022 \cdot 10^{23} mole^{-1} = 1,2044 \cdot 10^{25}$$

The pressure is then:

$$P = \frac{NKT}{V} = \frac{NKT}{V} = \frac{NKT}{Sh}$$

$$P = \frac{1,2044 \cdot 10^{25} \cdot 1,381 \cdot 10^{-23} \cdot 270\,K}{1000\,cm^2 \cdot 1\,m} = 4491\,hPa$$

Knowing that:

$$\begin{cases} P_i \cdot V = N \cdot K \cdot T_i \\ P_f \cdot V = N \cdot K \cdot T_f \end{cases}$$

We find the pressure after the 30°C rise:

$$P_f = \frac{P_i \cdot T_f}{T_i}$$

$$P_f = \frac{4491\,hPa \cdot 300\,K}{270\,K} = 4990\,hPa$$

Exercise 23

An iron container of mass 1 kg contains a mass of 3 kg of air.
If the initial temperature of the iron is 10°C and that of the air is 30°C, how much will the pressure in the container decrease once thermal equilibrium is reached?

The heat transferred from the air to the iron is the same as that absorbed by the iron.
At equilibrium we have:

$$T_{eq} = \frac{c_{s-aria} m_{aria} T_{i-aria} + c_{s-Fe} m_{Fe} T_{i-Fe}}{m_{aria} + m_{Fe}}$$

Knowing the specific heats of air and iron (respectively 0.72 and 440 J per kg and °C) we have:

$$T_{eq} = \frac{0,72 \frac{J}{kg°C} \cdot 3\,kg \cdot 30°C + 440 \frac{J}{kg°C} \cdot 1\,kg \cdot 10°C}{0,72 \frac{J}{kg°C} \cdot 3\,kg + 440 \frac{J}{kg°C} \cdot 1\,kg}$$

$$T_{eq} = \frac{4464,8\,J}{442,16 \frac{J}{°C}} = 10,1\,°C$$

The gas undergoes an isochoric transformation and therefore:

$$P_i V = NKT_i$$

$$P_f V = NKT_{eq}$$

From which:

$$\frac{P_f}{P_i} = \frac{(10,1 + 273,15)\,K}{(30 + 273,15)\,K} = 0,934$$

University-level exercises

Exercise 1

Given a mass M liters of water at the temperature of T (lower than saturation) contained in a closed system at pressure p, calculate the amount of heat Q that must be supplied to the system to make a fraction of the initial water vaporize at constant pressure .
What is the change in entropy between the final and initial states?
Draw the pV and TS diagram of the transformation.

Applying the first law of thermodynamics:

$$Q = M\left(h_2 - h_1\right) = M\left[\left(h_2 - h_{l,sat}(p_1)\right) - \left(h_{l,sat}(p_1) - h_1\right)\right] = M\left[x_2 r(p_1) + c_l(t_{sat}(p_1) - t_1\right]$$

Where latent heat and temperature are calculated at pressure pex is the fraction of the water to be vaporised.
The entropy change is given by:

$$\Delta S = M\left(s_2 - s_1\right) = M\left[\left(s_2 - s_{l,sat}(p_1)\right) - \left(s_{l,sat}(p_1) - s_1\right)\right] = M\left[x_2 \frac{r(p_1)}{t_{sat}(p_1)} + c_l \ln\left(\frac{t_{sat}(p_1)}{t_1}\right)\right]$$

The pV diagram is:

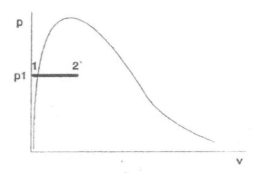

While that TS

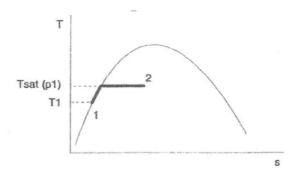

Exercise 2

In a room of size a, well the air is at a certain temperature.
Outside air has another temperature and a certain humidity.
If the title x of the external air coincides with the title of the internal air, calculate how much mass of water must be introduced into the room to reach the same humidity as the external one.

The title of the external air is:

$$x_e = 0.622 \frac{\varphi_e P_{sat}\left(t_e\right)}{P_{tot} - P_{sat}\left(t_e\right)}$$

The hygrometric degree of the air contained in the room is:

$$\varphi_{a,i} = \frac{P_{tot} x_e}{\left[(0.622 + x_e) P_{sat}(t_a)\right]}$$

The partial pressure of the air in the room is:

$$P_{a,i} = \varphi_{a,i} P_{sat}(t_a)$$

Applying the ideal gas equation of state, the mass of dry air in the room is:

$$M_a = \frac{P_a abh}{\dfrac{R_0}{m_{aria}}(273.15 + t_a)}$$

The title of the humid air in the room when it reaches the same humidity as the outside is:

$$x_{a,f} = 0.622 \frac{\varphi_e P_{sat}(t_a)}{P_{tot} - P_{sat}(t_a)}$$

From which the mass of water is obtained as:

$$M_v = M_a\left(x_{a,f} - x_{a,i}\right) = M_a\left(x_{a,f} - x_e\right)$$

Exercise 3

A system performs this thermodynamic step:

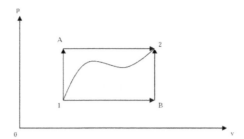

Moving from state 1 to state 2 along A, it absorbs 50 kcal of heat and does 20 kcal of work.
If, on the other hand, B follows, the heat absorbed is 36 Kcal.
What is the work along B?
If returning from 2 to 1 along the curved line, work is -13Kcal, what is Q?

According to the first law of thermodynamics, the change in internal energy is:

$$\Delta U12 := Q - L$$

$$\Delta U12 = 30\,kcal$$

Since the internal energy is a function of state we have:

$$L1B2 := Q1B2 - \Delta U12$$

$$L1B2 = 6\,kcal$$

$$L1B2 = 25.121kJ$$

Returning from 2 to 1 we have:

$$L21 := -13kcal$$

$$\Delta U21 := -\Delta U12$$

$$\Delta U21 = -125.604kJ$$

And then:

268

$$Q21 := \Delta U21 + L21$$

$$Q21 = -43\,kcal$$

$$Q21 = -180.032kJ$$

Exercise 4

A 4.57 kg system undergoes a polytropic transformation of exponent 1.35 in the initial state 1 with a pressure of 3.54 atm and a specific volume of 0.242 cubic meters per kilo which brings it to state 2 with a pressure of 1.88 atm.
Determine the final specific volume and the work done.

We have:

$$M := 4.57kg$$

$$n := 1.35$$

$$p1 := 3.54atm$$

$$v1 := 0.242\frac{m^3}{kg}$$

$$p2 := 1.88atm$$

From the polytropic equation we get:

$$v2 := v1 \cdot \left(\frac{p1}{p2}\right)^{\frac{1}{n}}$$

$$v2 = 0.387\frac{m^3}{kg}$$

The work done is:

$$L12 := M \cdot \frac{(p1 \cdot v1)}{n-1} \cdot \left[1 - \left(\frac{p2}{p1} \right)^{\frac{(n-1)}{n}} \right]$$

$$L12 = 171.507 \text{kJ}$$

Exercise 5

In a direct cycle the following quantities of heat are exchanged:

+3.56 x 10^2 kcal; + 4.28 x 10^2 kcal; -5.20 x 10^2 kcal; +0.834 x 10^2 kcal.

Determine cycle work and efficiency.

For a cycle, the variation of internal energy is zero, therefore, from the first principle, we have Q=L, i.e.:

$$L := (3.56 + 4.28 - 5.2 + 0.834) \cdot 10^2 \text{kcal}$$

$$L = 347.4 \text{kcal}$$

$$L = 1.454 \times 10^6 \text{J}$$

The yield is:

$$Qass := (3.56 + 4.28 + 0.834) \cdot 10^2 \text{kcal}$$

$$\eta := \frac{L}{Qass}$$

$$\eta = 0.401$$

Exercise 6

A container with rigid, fixed and adiabatic walls contains 202 grams of a fluid at a temperature of 60°C.
A stirrer delivers 3.2 KJ to the fluid.
Calculate the entropy change of the fluid assuming that:

$$c_V = cost = 0.71 \ kJ/kg \ K.$$

Since the system is adiabatic:

$$\delta q = 0$$

And by the first principle we have:

$$du = -\delta l$$

$$\Delta u = \int_1^2 c_v dT = -\frac{Le}{m}$$

We have:

$$c_v \Delta T = -\frac{Le}{m} \quad \rightarrow \quad \Delta T = -\frac{Le}{mc_v} = 22.3 \quad °C$$

So the fluid temperature is:

$$T_2 = 82.3 \ °C = 355.45 \ K$$

The entropy change is calculated as:

$$\int_1^2 dS = \Delta S_{1-2} = m \int_1^2 \frac{c_v dT}{T}$$

From which:

$$\int_1^2 dS = \Delta S_{1-2} = m\int_1^2 \frac{c_v dT}{T} = mc_v\int_1^2 \frac{dT}{T} = mc_v \ln\frac{T_2}{T_1} =$$

$$0.202 \times 0.71 \times \ln\frac{355.45}{333.15} = 9.29 \times 10^{-3} \frac{kJ}{K}$$

Exercise 7

A heat engine evolves according to a reversible Carnot cycle.
For each cycle a work of 40 KJ is obtainable.
If the efficiency is 0.35 and the cold source is at 40°C, calculate the temperature of the hot source, the quantities of heat exchanged and the entropy variation of the two sources.

We have:

$$L := 40 kJ$$

$$\eta := 0.35$$

$$T2 := (40 + 273.15) K$$

From the Carnot cycle efficiency expression:

$$T1 := \frac{T2}{1 - \eta}$$

$$T1 = 481.769 K$$

The absorbed heat will be:

$$Qass := \frac{L}{\eta}$$

$$Qass = 114.286 kJ$$

While the heat transferred will be:

$$Qced := L - Qass$$

$$Qced = -7.429 \times 10^4 \, J$$

The entropy variations of the two sources are:

$$\Delta Sc := \frac{-Qass}{T1}$$

$$\Delta Sc = -0.237 \frac{1}{K} \, kJ$$

$$\Delta Sf := \frac{-Qced}{T2}$$

$$\Delta Sf = 0.237 \frac{1}{K} \, kJ$$

Exercise 8

In a horizontal cylinder there is air at 20°C and 60 atm absolute.
The volume is 0.1 cubic meters.
With the following transformations: isobaric, isotherm, adiabatic, polytropic of exponent 1.5 the volume is doubled.
Calculate the significant parameters in the various cases.

We have:

$$V_1 = \frac{R \, T_1}{p_1} = \frac{287 \quad 293.15}{98100 \quad 60} = 1.4294 \; 10^{-2} \, \frac{m^3}{kg}$$

From which:

$$m = \frac{V_1}{v_1} = 6.99 \ kg$$

$$v_2 = \frac{V_2}{m} = 4.2918 \ 10^{-2} \ \frac{m^3}{kg}$$

For the isobar transformation:

$$\frac{V_1}{R \, T_1} = \frac{V_2}{R \, T_2} \quad \rightarrow \quad T_2 = \frac{V_2 \, T_1}{V_1} = 879 \ K$$

$$\Delta H_{12} = m \ \Delta h_{12} = m \ c_p \ \Delta T_{12} = 6.99 \ 1.005 \ (879 - 293) = 4116.62 \ kJ$$

$$\Delta U_{12} = m \ \Delta U_{12} = m \ c_v \ \Delta T_{12} = 2936.9 \ kJ$$

$$Q_{12} = m \ c_p \ \Delta T_{12} = 6.99 \ 1.005 \ (879 - 293) = 4116.62 \ kJ$$

$$L_{12} = Q_{12} - \Delta U_{12} = 1179.72 \ kJ$$

$$\Delta S_{12} = m \ \Delta s_{12} = m \ c_p \ \ln \frac{T_2}{T_1} = 7.717 \ \frac{kJ}{K}$$

For the isotherm:

$$p_2 = p_1 \frac{V_1}{V_2} = 1.962 \ MPa$$

$$\Delta U_{12} = 0 \qquad e \qquad \Delta H_{12} = 0$$

$$L_{12} = m \int_1^2 p \ dv = m \int_1^2 \frac{RT}{V} \ dv = m \ R \ T \ \ln \frac{V_2}{V_1} = 645 \ kJ$$

$$Q_{12} = L_{12} = 645 \ kJ$$

$$\Delta S = \frac{Q_{12}}{T} = 2.201 \ \frac{kJ}{K}$$

For the adiabatic:

$$T_2 = T_1 \left(\frac{V_1}{V_2} \right)^{k-1} = 188.81 \ K$$

$$p_1 = p_2 \left(\frac{V_2}{V_1} \right)^{k} = 1.263 \ MPa$$

$$\Delta H_{12} = m \, \Delta h_{12} = m \, c_p \, \Delta T_{12} = 6.99 \ 1.005 \ (188.81 - 293) = -731 \ kJ$$

$$\Delta U_{12} = m \, \Delta U_{12} = m \, c_v \, \Delta T_{12} = -522.23 \ kJ$$

$$L_{12} = -\Delta U_{12} = 522.23 \ kJ$$

$$\Delta S_{12} = 0$$

For polytropics:

$$c_n = c_v \frac{n-k}{n-1} = 0.1434 \frac{kJ}{kg \ K}$$

$$p_2 = p_1 \left(\frac{V_1}{V_2} \right)^{n} = 60 \ 9.81 \ 10^4 \left(\frac{1.4294 \ 10^{-2}}{4.2918 \ 10^{-2}} \right)^{1.5} = 1.132 \ MPa$$

$$T_2 = \frac{p_2 \, V_2}{R} = 169 \ K$$

$$Q_{12} = m \, c_n \, \Delta T_{12} = -124.43 \ kJ$$

$$\Delta U_{12} = m \, \Delta U_{12} = m \, c_v \, \Delta T_{12} = -622.356 \ kJ$$

$$L_{12} = Q_{12} - \Delta U_{12} = 497.92 \ kJ$$

Exercise 9

Given oxygen, assuming an ideal gas with k=1.4, we have the following cycle:

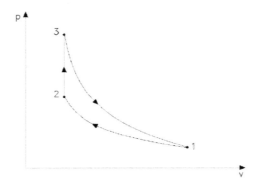

Point 1 is at 0.9 bar and 0.88 cubic meters per kg.

Point 3 has 21.5 bars.

The transformation 1-2 is isothermal, the 2-3 isochoral, the 3-1 polytropic of exponent 1.32

Determine the maximum and minimum temperature of the cycle, the quantity of heat exchanged along the individual transformations and the efficiency of the cycle.

The constant R for oxygen is:

$$R = \frac{R^*}{M_{O_2}} = \frac{8314}{32} = 259.8 \frac{J}{kg\ K}$$

Therefore:

$$T_1 = \frac{p_1 V_1}{R} = \frac{0.9 \cdot 10^5 \cdot 0.88}{259.8} = 304.8\ K$$

$$T_2 = T_1 = 304.8\ K$$

From the polytropic we obtain:

$$T_3 = T_1 \left(\frac{p_3}{p_1} \right)^{\frac{n-1}{n}} = 304.8 \left(\frac{21.5}{0.9} \right)^{\frac{0.32}{1.32}} = 657.8\ K$$

And then:

$$V_3 = \frac{RT_3}{p_3} = \frac{259.8 \cdot 657.8}{21.5 \cdot 10^5} = 79.5 \cdot 10^{-3} \frac{m^3}{kg} = V_2$$

$$p_2 = p_3 \frac{T_2}{T_3} = 21.5 \frac{304.8}{657.8} = 9.96 \ bar = 996 \ kPa$$

For the isotherm, the heat exchanged is:

$$q_{12} = l_{12} = RT_1 \ln \frac{V_2}{V_1} = 259.8 \cdot 304.8 \cdot \ln \left(\frac{79.5 \cdot 10^{-3}}{0.88} \right) = -190.4 \frac{kJ}{kg}$$

For the isochore it holds:

$$q_{23} = c_v (T_3 - T_2) = \frac{R}{k-1}(T_3 - T_2) = \frac{259.8}{1.4-1}(657.8 - 304.8) = 229.3 \frac{kJ}{kg}$$

While for polytropics:

$$q_{31} = c_n (T_1 - T_3) = \frac{R}{k-1}\frac{k-n}{1-n}(T_1 - T_3) = \frac{259.8}{1.4-1}\frac{1.4-1.32}{1-1.32}(304.8 - 657.8) = 57.8 \frac{kJ}{kg}$$

The work per unit mass is:

$$l = q_{12} + q_{23} + q_{31} = -190.4 + 229.3 + 57.3 = 96.2 \frac{kJ}{kg}$$

The yield is therefore:

$$\eta = \frac{l}{q_{23} + q_{31}} = \frac{96.2}{229.3 + 57.3} = 0.336$$

Exercise 10

The specific heat flux collected by a solar collector is 600 W per square meter when the collector operates at 90°C.

If the air is at a temperature of 21°C, calculate the minimum area to supply 1 kW.

Minimum air occurs when efficiency is maximum, i.e. when the collector works like a Carnot machine.
The maximum achievable yield is:

$$\eta = 1 - \frac{T_0}{T} = 1 - \frac{21 + 273.15}{90 + 273.15} = 0.19$$

So the required thermal power is:

$$q = \frac{P}{\eta} = \frac{10^3}{0.19} = 5263 \ W$$

From which:

$$A = \frac{q}{\phi} = \frac{5263}{600} = 8.77 \ m^2$$

Exercise 11

In a circular tube with a diameter of 5.08 cm, humid water vapor flows at a temperature of 271°C and title x=0.98.
The maximum flow rate is 1'134 Kg/s.
Determine the speed assuming the mixture is homogeneous.

From the tables, linearly interpolating the values of the specific volumes we have, for water:

$$V_{l271} = V_{l270} + \frac{V_{l280} - V_{l270}}{280 - 270} (271 - 270) = 1305 \ \frac{cm^3}{kg} = 1.305 \ 10^{-3} \ \frac{m^3}{kg}$$

While for steam:

278

$$V_{v271} = 0.03501 \frac{m^3}{kg}$$

The biphasic mixture will have:

$$v = (1-x)\, V_{l271} + x\, V_{v271} = 0.03501 \frac{m^3}{kg}$$

So the speed will be:

$$w = \frac{4 v G}{\pi\, d^2} = 19.2 \frac{m}{s}$$

Exercise 12

A steam cycle with two superheats has the following parameters.
For the first turbine:

$$\mathbf{p_3 = 100\ bar,\ t_3 = 500\ ^\circ C,\ \eta_{\vartheta I} = 0.75,\ p_{3'} = 30\ bar}$$

For the booster heater:

$$\mathbf{p_{3''} = 30\ bar,\ t_{3''} = 500\ ^\circ C}$$

For the second turbine:

$$\eta_{\vartheta II} = 0.85,\ \mathbf{p_4 = 0.05\ bar.}$$

Calculate the efficiency of the system.

The pattern in a TS diagram is given by:

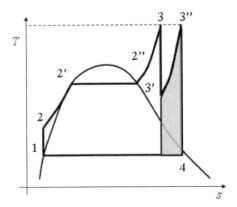

From the Mollier diagram, the enthalpy at point 3 is calculated:

$$h_3 = 3375 \quad \frac{kJ}{kg}$$

Point 3' bis has an enthalpy (calculated using the isentropic equation):

$$h_{3'bis} = 3030 \quad \frac{kJ}{kg}$$

From the efficiency of the first turbine we obtain:

$$h_{3'} = h_3 + \eta_{\vartheta I} \left(h_{3'bis} - h_3 \right) = 3116 \quad \frac{kJ}{kg}$$

The enthalpy of point 3" is:

$$h_{3''} = 3455 \quad \frac{kJ}{kg}$$

If we move along an isentropic line until we meet the isobar with pressure equal to p4 we have:

$$h_{4\,bis} = 2210 \ \frac{kJ}{kg}$$

The enthalpy of point 4 is obtained from the efficiency of the second turbine:

$$h_4 = h_{3^{\cdot}} + \eta_{\vartheta II} \left(h_{4bis} - h_{3^{\cdot}} \right) = 3455 + 0.85 \left(2210 - 3455 \right) = 2397 \ \frac{kJ}{kg}$$

From the tables of the saturated liquid we obtain, for point 1:

$$h_1 = 136 \ \frac{kJ}{kg}$$

The useful power is:

$$P_u = G_v \left[\left(h_3 - h_{3^{\cdot}} \right) + \left(h_{3^{\cdot \cdot}} - h_4 \right) \right]$$

The heat supplied to superheat is instead:

$$\dot{Q} = G_v \left[\left(h_3 - h_1 \right) + \left(h_{3^{\cdot \cdot}} - h_{3^{\cdot}} \right) \right]$$

And the efficiency of the whole system is given by:

$$\eta = \frac{P_u}{\dot{Q}} = \frac{\left(h_3 - h_{3^{\cdot}} \right) + \left(h_{3^{\cdot \cdot}} - h_4 \right)}{\left(h_3 - h_1 \right) + \left(h_{3^{\cdot \cdot}} - h_{3^{\cdot}} \right)} = 0.36$$

Exercise 13

A Joule cycle uses air as the driving fluid with:

$$c_p = 1.005 \ kJ/kg \ K; \ K = 1.4$$

$$R = 0.287 \ kJ/kg \ K)$$

And it is characterized by the following parameters:

$p_1 = 1$ **bar**; $t_1 = 10$ **°C** $= 283.15$ **K**; $p_3 = 5$ **bar**; $t_3 = 700$ **°C** $= 973.15$ **K**.

Determine the work of expansion and compression per kg of fluid and the cycle efficiency.

The cycle pattern in a TS diagram is:

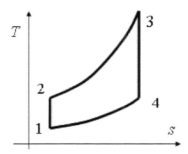

The temperature in 4 is obtained from:

$$T_4 = T_3 \left(\frac{p_4}{p_3} \right)^{\frac{K-1}{K}} = 973.15 \left(\frac{1}{5} \right)^{\frac{1.4-1}{1.4}} = 614.4 \quad K$$

And the expansion work is:

$$L_e = L_{34} = -(h_4 - h_3) = -c_p (T_4 - T_3) = -1.005 (614.4 - 973.15) = 360.5 \quad \frac{kJ}{kg}$$

The temperature at point 2 is:

$$T_2 = T_1 \left(\frac{p_2}{p_1} \right)^{\frac{k-1}{k}} = 448.46 \quad K$$

And the compression work is:

$$L_c = L_{12} = -(h_2 - h_1) = -c_p (T_2 - T_1) = -1.005 (448.46 - 283.15) = -166.1 \ \frac{kJ}{kg}$$

The cycle efficiency is:

$$\eta = \frac{L_u}{Q} = \frac{L_e + L_c}{c_p (T_3 - T_2)} = \frac{360.5 - 166.1}{1.005 (973.15 - 418.46)} = 0.37$$

Exercise 14

Calculate the reversible work done by a cylinder-piston system compressing 2 moles of pure nitrogen.
The data are:

$$V_1 = 30 \ l; \ V_2 = 10 \ l;$$

$$T_1 = T_2 = 300 \ K;$$

Peso atomico dell'azoto = 14;

$$T_c \ (azoto) = 126,27 \ K.$$

$$R_{N2} = 0,297 \ kJ/kgK;$$

$$a = 72,72 \ m^6 Pa/kg^2; \ b = 0,89*10^{-3} \ m^3/kg;$$

For a closed system the work is:

$$L_{1-2} = \int_1^2 P \, dv$$

In the case of an ideal gas:

$$Pv = RT \rightarrow P = \frac{RT}{v}$$

$$L_{12} = \int_1^2 P \, dv = \int_1^2 \frac{RT}{v} \, dv = RT \int_1^2 \frac{1}{v} \, dv = RT (\ln v_2 - \ln v_1) = RT \left(\ln \frac{v_2}{v_1} \right)$$

$$numero\ moli = = 2 \cdot 28 = 56\ g = 0,056\ kg$$

Therefore:

$$v_1 = \frac{0,03}{0,056} = 0,5357\ \frac{m^3}{kg}\ ; \qquad v_2 = \frac{0,01}{0,056} = 0,1786\ \frac{m^3}{kg}$$

$$L_{12} = RT \left(\ln\frac{v_2}{v_1} \right) = 0,297 \cdot 300 \cdot \left(\ln\frac{0,1786}{0,5357} \right) = -97,87\ \frac{kJ}{kg}$$

In the case of Van der Waals gas:

$$(v - b)\left(P + \frac{a}{v^2} \right) = RT \rightarrow P = \left(\frac{RT}{v - b} - \frac{a}{v^2} \right)$$

$$L_{12} = \int_1^2 P\,dv = \int_1^2 \left(\frac{RT}{v - b} - \frac{a}{v^2} \right) dv =$$

$$\int_1^2 \frac{RT}{v - b}\,dv - \int_1^2 \frac{a}{v^2}\,dv = RT \int_1^2 \frac{1}{v - b}\,dv - a \int_1^2 \frac{1}{v^2}\,dv =$$

$$= RT \ln\left(\frac{v_2 - b}{v_1 - b} \right) + a\left(\frac{1}{v_2} - \frac{1}{v_1} \right) = -98,44\ \frac{kJ}{kg}$$

Exercise 15

Assuming that, at thermal equilibrium with ice at atmospheric pressure, a rod of mercury measures 1 mm and at thermal equilibrium with the boiling point of water at atmospheric pressure, it measures 8 mm, determine a temperature scale.
Suppose further to immerse the rod in a fluid and at equilibrium to measure a length of 4.6 mm.

We have:

$$\Delta L = L_{finale} - L_{iniziale} = 8 - 1 = 7 \text{ mm}$$

$$\Delta T = T_{finale} - T_{iniziale} = 100 - 0 = 100 \text{ °C}$$

$$\Delta L / \Delta T = 7 / 100 = 0.07 \text{ mm/°C}$$

The 3.6 mm stretch corresponds to:

$$\Delta L / \Delta T = 0.07 \quad \longrightarrow \quad \Delta T = \Delta L / 0.07 = 3.6 / 0.07 = 51.43 \text{ °C}$$

Exercise 16

20 grams of hydrogen are compressed adiabatically at 300 K from 1 bar to 10 bar, with a cylinder-piston system whose piston has a surface area of 0.5 square metres.
Calculate reversible and irreversible work considering that these compressions take 300 and 60 seconds with reaction forces of 500 N and 800 N.
Hydrogen has the following data:

$R = 4,124 \ kJ/kgK;$

$\gamma_p = calore \ specifico \ a \ P \ costante \ (a \ 25 \ °C \ e \ bassa \ P) = 14,302 \ kJ/kgK;$

$\gamma_v = calore \ specifico \ a \ V \ costante \ (a \ 25 \ °C \ e \ bassa \ P) = 10,178 \ kJ/kgK.$

Reversible work is:

$$L_{rev} = \int_{A}^{B} P \ dv$$

The transformation is therefore adiabatic:

$$Pv^k = \text{cost} \rightarrow P = \frac{\text{cost}}{v^k}; k = \frac{\gamma_p}{\gamma_v} = \frac{14,302}{10,178} = 1,405$$

And then:

$$v_2 = \sqrt[k]{\frac{\text{cost}}{P_2}} = {}^{1.405}\sqrt{\frac{3426716}{1000000}} = {}^{1.405}\sqrt{3.43} = 2.404\frac{m^3}{kg}$$

$$L_{rev} = \int_A^B P\,dv = {} = -2875 \cdot 0.02 = -57.5\,kJ$$

The irreversible work is:

$$L_{irr} = L_{rev} + \int R \cdot dx$$

Where R is given by (in both cases):

$$R_1 = 500 \text{ N}; \ R_2 = 800 \text{ N}$$

Therefore:

$$L_{irr} = L_{rev} + \int R \cdot dx = L_{rev} + R \cdot (x_2 - x_1) = L_{rev} + R \cdot \Delta x$$

But:

$$\Delta V = \Delta v \cdot m = (v_2 - v_1) \cdot m = (2.404 - 12.372) \cdot 0.02 = -0.199\,m^3$$
$$\Delta x = \frac{\Delta V}{S} = -\frac{0.199}{0.5} = -0.398\,m$$

In both cases we have:

$$L_{irr} = L_{rev} + R \cdot \Delta x \rightarrow L_{irr(1)} = L_{rev} + R_1 \cdot \Delta x = -57500 - 500 \cdot 0.398 = -57699\,J$$
$$L_{irr(2)} = L_{rev} + R_2 \cdot \Delta x = -57500 - 800 \cdot 0.398 = -57818\,J$$

Exercise 17

Given the Diesel cycle:

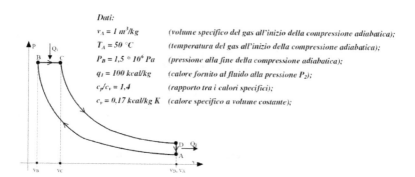

Dati:

$v_A = 1 \, m^3/kg$ (volume specifico del gas all'inizio della compressione adiabatica);

$T_A = 50 \, °C$ (temperatura del gas all'inizio della compressione adiabatica);

$P_B = 1,5 * 10^6 \, Pa$ (pressione alla fine della compressione adiabatica);

$q_1 = 100 \, kcal/kg$ (calore fornito al fluido alla pressione P_2);

$c_p/c_v = 1,4$ (rapporto tra i calori specifici);

$c_v = 0,17 \, kcal/kg \, K$ (calore specifico a volume costante);

Calculate the amount of heat released by the fluid and the efficiency of the cycle.

The heat and efficiency are:

$$q_2 = c_v(T_D - T_A)$$

$$\eta = 1 - \frac{q_2}{q_1} = 1 - \frac{c_v(T_D - T_A)}{c_p(T_C - T_B)}$$

Where is it:

$$T_A = 50 \, °C = 50 + 273 = 323 \, K.$$

We get the other temperatures.
The transformation AB is adiabatic therefore:

$$T_A P_A^{\frac{1-k}{k}} = T_B P_B^{\frac{1-k}{k}}$$

And then:

$$T_B = T_A \left(\frac{P_A}{P_B} \right)^{\frac{1-k}{k}}$$

Since:

$$c_p = k \cdot c_v = 1.4 \cdot 0.17 = 0.238 \frac{kcal}{kg \cdot K} = 238 \frac{cal}{kg \cdot K} = 238 \cdot 4.186 = 996.268 \frac{J}{kg \cdot K}$$

$$c_v = 0.17 \cdot 1000 \cdot 4.186 = 711.62 \frac{J}{kg \cdot K}$$

$$R = c_p - c_v = 996.268 - 711.62 = 284.65 \frac{J}{kg \cdot K}$$

We have:

$$P_A = \frac{RT_A}{v_A} = \frac{284.65 \cdot 323}{1} = 91942 \ Pa$$

$$T_B = T_A \left(\frac{P_A}{P_B}\right)^{\frac{1-k}{k}} = 323 \left(\frac{91942}{1.5 \cdot 10^6}\right)^{\frac{1-1.4}{1.4}} = 323 \cdot (0.0613)^{-0.2857} = 717.22 \ K$$

Moreover:

$$q_1 = 100 \frac{kcal}{kg} = 100 \cdot 1000 \cdot 4.186 = 418600 \frac{J}{kg}$$

$$q_1 = c_p (T_C - T_B) \quad \rightarrow \quad T_C = \frac{q_1}{c_p} + T_B = \frac{418600}{996.268} + 717.22 = 1137.4 \ K$$

In the end:

$$T_D = T_C \left(\frac{v_C}{v_D}\right)^{1.4-1} = 1137.4 \left(\frac{0.21586}{1}\right)^{0.4} = 616 \ K$$

The results sought are therefore:

$$q_2 = c_v (T_D - T_A) = 711.62 \cdot (616 - 323) = 208505 \frac{J}{kg} = 208.5 \frac{kJ}{kg}$$

$$\eta = 1 - \frac{q_2}{q_1} = 1 - \frac{c_v (T_D - T_A)}{c_p (T_C - T_B)} = 1 - \frac{711.62 \ (616 - 323)}{996.268 \ (1137.4 - 717.22)} = 1 - \frac{208505}{418612} = 0.502$$

Exercise 18

Consider an Otto-cycle engine with air as the working fluid and whose compression ratio is 10.

Knowing that 1 kg of fuel is consumed every 10 minutes, calculate the cycle efficiency and the mechanical power developed with the following data:

potere calorifico inferiore (= LHV = Lower Heating Value) = 10000 kcal/kg;
$k_{aria} = 1,4.$

The yield is calculated with the definition:

$$\eta_{OTTO} = 1 - \left(\frac{1}{\rho}\right)^{k-1} = 1 - \left(\frac{1}{10}\right)^{1,4-1} = 0,602$$

The calorific value is:

$$LHV = 10000\frac{kcal}{kg} = 10000000\frac{cal}{kg} = 10000000 \cdot 4,186 = 41860000\frac{J}{kg} = 41860\frac{kJ}{kg}$$

In 10 minutes you will have:

$$Q_1 = 41860 \cdot 1 = 41860 \, kJ$$
$$t = 10 \cdot 60 = 600 \, s$$

The thermal power is:

$$\dot{Q}_1 = \frac{41860000 \cdot 1}{10 \cdot 60} = 69767 \, \frac{J}{s} = 69,8 \, kW$$

The mechanical power will be:

$$W = \dot{Q}_1 \, \eta = 69,8 \cdot 0,602 = 42 \, kW$$

Exercise 19

One plant operates on the Brayton cycle providing 20,000 HP.
The maximum and minimum temperatures are 840 °C and 16 °C.
The maximum and minimum pressures are 4 bar and 1 bar.
Calculate the power developed by the turbine and the required mass flow
with the following data:

$$k = 1,4$$

$$R = 0,287 \ kJ/kgK$$

The cycle in the pV and TS diagram is:

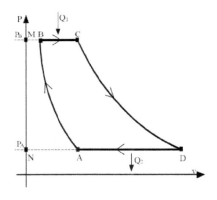

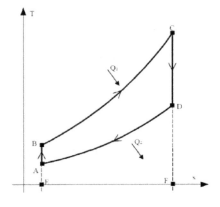

The work produced by the turbine is:

$$L_A = L_T - L_C$$

The work required by the compressor is:

$$L_C = h_B - h_A = c_P(T_B - T_A)$$

The temperature at B is given by:

$$T_B = T_A \cdot \left(\frac{P_A}{P_B}\right)^{\frac{1-k}{k}} = T_A \cdot \left(\frac{P_B}{P_A}\right)^{\frac{k-1}{k}} = 289 \cdot \left(\frac{4}{1}\right)^{\frac{1.4-1}{1.4}} = 429,4K$$

The specific heat is obtained from:

$$\cdot\begin{cases} \frac{c_P}{c_v} = k \\ c_P - c_v = R \end{cases} \rightarrow \begin{cases} c_P = k \cdot c_v \\ k \cdot c_v - c_v = R \end{cases} \rightarrow \begin{cases} - \\ c_v(k-1) = R \end{cases} \rightarrow \begin{cases} - \\ c_v = \frac{R}{(k-1)} \end{cases} \rightarrow \begin{cases} c_P = k\frac{R}{(k-1)} \\ c_v = \frac{R}{(k-1)} \end{cases}$$

Therefore:

$$L_C = h_B - h_A = c_P(T_B - T_A) = k\frac{R}{(k-1)} \cdot (T_B - T_A) = 1,4\frac{0,287}{0,4}(429,4 - 289) = 141\frac{kJ}{kg}$$

Moreover:

$$T_D = T_C \cdot \left(\frac{P_C}{P_D}\right)^{\frac{1-k}{k}} = 1113 \cdot \left(\frac{4}{1}\right)^{\frac{1-1,4}{1,4}} = 749\ K$$

$$L_T = h_C - h_D = c_P(T_C - T_D) = k\frac{R}{(k-1)} \cdot (T_C - T_D) = 1,4\frac{0,287}{0,4}(1113 - 749) = 365,6\frac{kJ}{kg}$$

The net work is:

$$L_A = L_T - L_C = 365,6 - 141 = 224,6\ \frac{kJ}{kg}$$

The mass flow will be:

$$G_M = \frac{W_A}{L_A} = \frac{20000}{1,36} \cdot \frac{1}{224,6} = 65,47 \frac{kg}{s}$$

And the power developed by the turbine is:

$$W_T = G_M \cdot L_T = 65,47 \cdot 365,6 = 24275,8 \; kW = 32554 \; CV$$

Exercise 20

Given a Rankine cycle covered by water with an evaporation temperature in the boiler of 250°C and a condensing temperature of 20°C.
Calculate the efficiency of the cycle, the work produced by the cycle per unit of mass, the ideal mechanical power developed in the turbine with a flow rate of 2,000 kg/h.
The data are:

x_3 = titolo del vapore d'acqua all'ingresso in turbina = 1;

x_4 = titolo del vapore d'acqua alla fine dell'espansione in turbina = 0,7;

c_p = calore specifico dell'acqua (allo stato liquido) = 1 kcal/kg°C;

r_{T1} = calore di evaporazione alla temperatura di condensazione = 2453 kJ/kg;

r_{T2} = calore di evaporazione alla temperatura di evaporazione = 1715 kJ/kg.

The cycle TS diagram is:

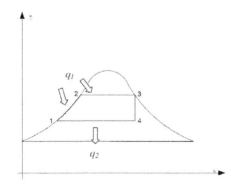

The yield is:

$$\eta = 1 - \frac{q_2}{q_1} = 1 - \frac{r_{T_1} \cdot x_4}{c_p(T_2 - T_1) + r_{T_2} \cdot x_3}$$

Where is it:

$$q_2 = r_{T_1} \cdot x_4 = 2453 \cdot 0,7 = 1717,1 \frac{kJ}{kg}$$

$$q_1 = c_p(T_2 - T_1) + r_{T_2} \cdot x_3 = 4,186 \cdot (250 - 20) + 1715 \cdot 1 = 962,78 + 1715 = 2677,8 \frac{kJ}{kg}$$

Therefore:

$$\eta = 1 - \frac{q_2}{q_1} = 1 - \frac{1717,1}{2677,8} = 0,359$$

The work produced is:

$$L = q_1 - q_2 = 2677,8 - 1717,1 = 960,7 \frac{kJ}{kg}$$

While the power will be:

$$W = L \cdot G_M = 5337 \, kW$$

Exercise 21

Consider a wall formed by a layer of plaster of 1 cm, a layer of bricks of 14 cm and another layer of plaster of 1 cm.
Plaster has a thermal conductivity of 0.29 W/mK, bricks 0.5 W/mK.
The outside temperature is -5 °C, the inside 20 °C.
Calculate the heat loss per unit area through the wall assuming that the convection coefficient is 7 W/mKm.
If we wanted to reduce the waste heat by 30%, how much insulation thickness should we put if its conductivity is 0.02 W/mK.

In the absence of insulation, the overall thermal resistance is:

$$R_{th,tot} = \frac{1}{\alpha} + \frac{s_1}{\lambda_1} + \frac{s_2}{\lambda_2} + \frac{s_3}{\lambda_3} + \frac{1}{\alpha}$$

And it is worth 0.6346.
The heat dissipated is:

$$q = \frac{\left(t_i - t_e\right)}{R_{th,tot}}$$

And it's worth 39.39W per square metre.
With the insulator, this heat must be:

$$q^* = q(1 - r)$$

That is 27,573 W per square metre.
The insulation thickness will be:

$$s_{is} = \left(\frac{\left(t_i - t_e\right) - q * R_{th,tot}}{q * / \lambda_{is}} \right)$$

And it will be 5.44cm.

Exercise 22

A wall consists of a layer of masonry of 30 cm with plaster of 2.5 cm on both faces (thermal conductivity of plaster 1.2 W/mK and of masonry 1.45 W/mK).
It is 4°C outside and 19°C inside.
The convection coefficients are 10 W/mKm and 20 W/mKm respectively indoors and outdoors.
Calculate the heat flux.

We have the case schematized in the figure:

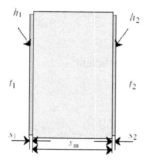

Considering the various contributions we will have:

$$K = \cfrac{1}{\cfrac{1}{h_1}+\cfrac{s_1}{k_1}+\cfrac{s_m}{k_m}+\cfrac{s_2}{k_2}+\cfrac{1}{h_2}} = \cfrac{1}{\cfrac{1}{10}+\cfrac{2.5\ 10^{-2}}{1.2}+\cfrac{0.3}{1.45}+\cfrac{2.5\ 10^{-2}}{1.2}+\cfrac{1}{20}} = 2.509 \ \ \frac{W}{m^2\ {}^\circ C}$$

And so the heat flux is:

$$\dot{q}_{12} = K(t_1 - t_2) = 2.509\,(19-4) = 37.6 \ \ \frac{W}{m^2}$$

Exercise 23

Given a piece of steel immersed in an air environment at 100°C, there is the presence of a convection current along the wall of the piece.
Two thermocouples placed 20 mm and 10 mm from the surface reveal temperatures of 20 °C and 25 °C.
Calculate the convection coefficient between the air and the wall.

Knowing the thermal conductivity value of the steel, the heat flux will be:

$$q = k\frac{(T_2 - T_1)}{(s_1 - s_2)} = 44.7 \left(\frac{4186}{3600}\right)\frac{(25-20)}{(0.02-0.01)} = 25.9 \ \ \frac{kW}{m^2}$$

The wall temperature will be:

295

$$T_p = T_1 + q\frac{S_1}{k} = 25 + 25900 \ \frac{0.01}{44.7 \left(\dfrac{4186}{3600}\right)} = 30 \quad °C$$

From which we get the desired coefficient:

$$\begin{cases} q_{conv} = q \\ h = \dfrac{q}{(T_a - T_p)} = \dfrac{25900}{(100 - 30)} = 371.4 \quad \dfrac{W}{m^2 K} \end{cases}$$

Exercise 24

A steel pipe is traversed by water vapor at a pressure of 300 KPa and a temperature of 230°C.
The tube has an external diameter of 108 mm and a thickness of 3.75 mm and is placed in air at a temperature of 37 °C.
The pipe is covered with an insulating layer.
It is known that the thermal conductivity of the steel and of the insulation are 75 and 0.055 while the convective coefficients at the internal and external walls are 50 and 10.
If the external surface must not exceed 62°C, what will be the thickness of the insulation?

We have the following scheme:

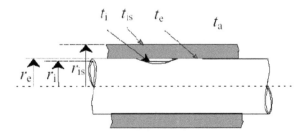

The transmittance per unit length is given by:

$$K = \cfrac{2\pi}{\cfrac{1}{h_i\, r_i} + \cfrac{\log\left(\cfrac{r_e}{r_i}\right)}{k_t} + \cfrac{\log\left(\cfrac{r_{is}}{r_e}\right)}{k_{is}} + \cfrac{1}{h_e\, r_{is}}}$$

And the heat flux is:

$$\dot{q}_L = K(t_f - t_a)$$

But it must also hold:

$$\dot{q}_L = h_e\,(t_{is} - t_a)$$

And then:

$$t_{is} = t_a + \frac{\dot{q}_L}{h_e} = t_a + \frac{k\left(t_f - t_a\right)}{h_e} \le t_m$$

From which:

$$K \le h_e\,\frac{t_m - t_a}{t_f - t_a} = 10\,\frac{62-37}{230-37} = 1.295 \quad \frac{W}{m\,K}$$

Substituting in the previous expression:

$$\frac{0.1}{r_{is}} + 18.1818 \ \log\left(r_{is}\right) \ge -48.616$$

$$r_{is} \ge 0.063 \quad m$$

The insulation must have a minimum thickness of:

$$e = r_{is} - r_e \ge 63 - 54 = 9 \quad mm$$

Exercise 25

A fluid stream runs parallel to a rectangular wall with sides 10 cm and 30 cm.
The temperature of the wall is 60°C while that of the fluid is 10°C.
The velocity of the fluid is 0.4 m/s.
Calculate the heat flow if the fluid is air at atmospheric pressure or if it is water.

We observe that for laminar motion, the Nusselt number will be:

$$Nu_L = \frac{h\,L}{k} = 0.664\ \mathrm{Re}_L^{1/2}\ \mathrm{Pr}^{1/3}$$

For the air we have a temperature of 308 K and from the tables we obtain:

$$k = 26.85\quad 10^{-3}\quad \frac{W}{m\,K}$$

$$\mathrm{Pr} = 0.705$$

$$v = 16.44\quad 10^{-6}\quad \frac{m^2}{s}$$

$$\mathrm{Re}_L = \frac{w\,L}{v} = \frac{0.4\ \ 0.1}{16.44\ \ 10^{-6}} = 2433$$

The motion is really laminar and therefore:

$$h = \frac{k}{L}Nu_L = \frac{k}{L}0.664\ \mathrm{Re}_L^{1/2}\ \mathrm{Pr}^{1/3} = 7.83\quad \frac{W}{m^2\,K}$$

Hence the heat exchanged is:

$$\dot{Q}' = h\left(t_p - t_f\right)b\,L = 7.83\left(60 - 10\right)\ 0.3\ \ 0.1 = 11.7\quad W$$

For water the temperature will be 35 °C and therefore:

$$k = 0.625 \quad \frac{W}{m\,K}$$

$$\mathrm{Pr} = 4.80$$

$$v = 7.22 \;\; 10^{-7} \quad \frac{m^2}{s}$$

$$\mathrm{Re}_L = \frac{w\,L}{v} = \frac{0.4 \;\; 0.1}{7.22 \;\; 10^{-7}} = 5.5 \;\; 10^4$$

Also in this case the motion is laminar and therefore:

$$h = \frac{k}{L}\,Nu_L = \frac{k}{L}\,0.664 \;\; \mathrm{Re}_L^{1/2}\,\mathrm{Pr}^{1/3} = 1647 \quad \frac{W}{m^2\,K}$$

The heat flux will be:

$$\dot{Q}'' = h\left(t_p - t_f\right) b\,L = 1647\,(60-10)\;\; 0.3 \;\; 0.1 = 2.47 \quad kW$$

Exercise 26

Water flows in a black plastic tube at a pressure of 0.1 Kg/s.
The tube has a diameter of 60 mm and a negligible thickness.
Water enters the pipe at 20°C.
The pipe is exposed to sunlight and receives a uniform heat flux of 2 kW per square metre.
How long must the pipe be for the water to come out at 80°C?

Note that only half of the top tube is exposed to the sun and will be heated. By applying the first law of thermodynamics we have that the heat absorbed by radiation will be the same as that needed to raise the temperature of the water:

$$\dot{q}\frac{\pi D}{2}\,L = \dot{m}\,c_p\,(T_2 - T_1)$$

So the length of the tube is:

$$L = \frac{2\dot{m}\,c_p}{\dot{q}\pi D}(T_2 - T_1) = \frac{0.1\ \ 4181}{2000\ \ 3.14\ \ 0.06}(80 - 20) = 66.58 \quad m$$

Exercise 27

A cryogenic fluid flows inside a long cylindrical tube with an external diameter of 20 mm and an external temperature of 77 K.
The external surface can be considered as a gray body of parameter 0.02.
The tube is contained in another cylindrical glass tube with an internal diameter of 50 mm.
The glass tube can be considered as a gray body at a temperature of 20 °C and parameter 0.08.
Calculate the thermal power exchanged by radiation between the two surfaces for each meter of pipe assuming that there is a vacuum inside.
If a cylindrical shield is placed between the two surfaces, coaxial to the two surfaces, with a diameter of 35 mm and a parameter of 0.02, by how much is the power exchanged per unit of length reduced in percentage terms?

In the absence of a screen, the thermal power is given by:

$$\frac{\dot{Q}}{L} = \frac{\sigma \pi d_1 \left(T_1^4 - T_2^4\right)}{\frac{1}{a_1} + \left(\frac{1-a_2}{a_2}\right)\frac{d_1}{d_2}} = \frac{5.67\ \ 10^{-8}\ \ 0.02\ \ \left(77^4 - 293^4\right)}{\frac{1}{0.02} + \left(\frac{1-0.08}{0.08}\right)\frac{0.02}{0.05}} = -0.152 \quad \frac{W}{m}$$

Having applied the Stefan-Boltzmann law.
If the screen is present, this further relationship will be obtained:

$$\left(\frac{Q}{L}\right)_{sch} = \frac{\sigma \pi d_1 \left(T_1^4 - T_3^4\right)}{\frac{1}{a_1} + \left(\frac{1-a_3}{a_3}\right)\frac{d_1}{d_3}} = \frac{\sigma \pi d_3 \left(T_3^4 - T_2^4\right)}{\frac{1}{a_3} + \left(\frac{1-a_2}{a_2}\right)\frac{d_3}{d_2}}$$

The unknown is obviously the temperature of the screen.
If we define:

$$\alpha_{12} = \frac{1}{a_1} + \left(\frac{1-a_2}{a_2}\right)\frac{d_1}{d_2} = \frac{1}{0.02} + \left(\frac{1-0.08}{0.08}\right)\frac{0.02}{0.05} = 54.6$$

$$\alpha_{13} = \frac{1}{a_1} + \left(\frac{1-a_3}{a_3}\right)\frac{d_1}{d_3} = \frac{1}{0.02} + \left(\frac{1-0.02}{0.02}\right)\frac{0.02}{0.035} = 78$$

$$\alpha_{32} = \frac{1}{a_3} + \left(\frac{1-a_2}{a_2}\right)\frac{d_3}{d_2} = \frac{1}{0.02} + \left(\frac{1-0.08}{0.08}\right)\frac{0.035}{0.05} = 58.05$$

So at equilibrium we have:

$$T_3 = \sqrt[4]{\frac{\dfrac{d_1 T_1^4}{\alpha_{13}} + \dfrac{d_3 T_2^4}{\alpha_{32}}}{\dfrac{d_1}{\alpha_{13}} + \dfrac{d_3}{\alpha_{32}}}} = \sqrt[4]{\frac{\dfrac{0.02\ 77^4}{78} + \dfrac{0.035\ 293^4}{58.05}}{\dfrac{0.02}{78} + \dfrac{0.035}{58.05}}} = 268.3 \quad K$$

From which we obtain the thermal power:

$$\left(\frac{Q}{L}\right)_{sch} = \frac{\sigma\pi d_1\left(T_1^4 - T_3^4\right)}{\dfrac{1}{a_1} + \left(\dfrac{1-a_3}{a_3}\right)\dfrac{d_1}{d_3}} = \frac{5.67\ 10^{-8}\ 0.02\ \left(77^4 - 268.3^4\right)}{\dfrac{1}{0.02} + \left(\dfrac{1-0.02}{0.02}\right)\dfrac{0.02}{0.035}} = -0.0748 \quad \frac{W}{m}$$

The percentage change will be:

$$\Delta = \frac{\left(\dfrac{Q}{L}\right)_{sch} - \dfrac{Q}{L}}{\dfrac{Q}{L}} 100 = 50.8\%$$

10

STATISTICAL PHYSICS

Equipartition principle

With the study of thermodynamics and heat transmission it was possible to lay the foundations for a new discipline given by statistical physics.

The very idea of applying statistical concepts to physics was completely new, with the exception of some ideas that had already appeared in kinematics, such as that of average speeds and accelerations, which had given rise to the first nucleus of statistical mechanics.

The kinetic theory of gases that we are going to explain represents a notable extension of the field of application of statistics in physical theories and therefore it can be said that statistical physics as a discipline took shape precisely with these studies.

The principle of equipartition of energy states that for each quadratic degree of freedom that makes up the motion of a system there is an associated energy equal to:

$$E_i = \frac{1}{2} kT$$

Where the Boltzmann constant and the absolute temperature appear

In classical thermodynamics, the energy contribution takes into account the kinetic energy of the particles (translational energy), the rotational energy due to the rotation of the particles around their center of gravity and the vibrational energy generated by the oscillation along the equilibrium axis.

Degrees of freedom are defined by the ways in which a system can move without constraints.

Said n the number of atoms that make up a molecule, the number of degrees of freedom for an ideal gas with linear molecules is given by 6n-5, while if the molecules are non-linear it is equal to 6n-6.

Kinetic theory of gases

The equipartition theorem derives from the application of Newtonian mechanics to the kinetic theory of gases, which has its foundations in statistics.

This theory describes a gas as a compound formed by a very large number of particles which are in constant random motion.

The particles, colliding with each other and hitting the walls of the container, give rise to detectable physical phenomena such as pressure, temperature and volume.

These studies went beyond (and at the same time denied) what Newton had said about the nature of pressure as a repulsive effect of particles.

Applying this theory, the statistical function of the equilibrium velocity distribution can be separated into the individual spatial terms:

$$P(v_x, v_y, v_z)dv_x dv_y dv_z = P(v_x)dv_x \cdot P(v_y)dv_y \cdot P(v_z)dv_z$$

$$P(v_i)dv_i = \frac{1}{\alpha\sqrt{\pi}} \exp\left(-\frac{v_i^2}{\alpha^2}\right)dv_i$$

where α is the value of the most probable speed that a particle can assume and is expressed as follows:

$$\alpha = \sqrt{\frac{2kT}{m}}$$

The mean speed and root mean square speed are derived as follows:

$$\langle v \rangle = \frac{2\alpha}{\sqrt{\pi}}$$

$$\langle v^2 \rangle = \frac{3}{2}\alpha^2$$

These quantities are also called thermal velocity and quadratic thermal velocity and directly correlate the velocity with the temperature.

The distribution of the modes as a function of the absolute values of the speeds is easily obtained and holds:

$$\frac{dN}{dv} = \frac{4N}{\alpha^3 \sqrt{\pi}} v^2 \exp\left(-\frac{mv^2}{2kT}\right)$$

Once s is defined as the number of degrees of freedom, the energy distribution takes the form:

$$P(E)dE = \frac{c(s)}{kT}\left(\frac{E}{kT}\right)^{\frac{1}{2}s-1} \exp\left(-\frac{E}{kT}\right)dE$$

Where c(s) depends on whether you have odd or even degrees of freedom. In the two cases mentioned, this factor is respectively:

$$s = 2l \Rightarrow c(s) = \frac{1}{(l-1)!}$$

$$s = 2l+1 \Rightarrow c(s) = \frac{2^l}{\sqrt{\pi}(2l-1)!}$$

This energy distribution is defined as the Maxwell-Boltzmann distribution and was enunciated by the latter in 1871, reworking the previous concepts of the distribution of speeds, in turn published in 1859 by Maxwell.
Almost all of the kinetic theory of gases was developed in the mid-nineteenth century, even if the first ideas date back to a century earlier, with the studies of Bernoulli.
This distribution is based on various hypotheses, among which there is the distinguishability of the particles, the linearity and the isotropy of the system.
Furthermore, the statistical processes underlying the state of the system correspond to Markov's laws of statistics.
Such a system is said to be perfectly thermalized and corresponds to the reality of gases if the frequency of collisions between the particles is high enough compared to the observation time.
The number of molecules that collide, in a certain amount of time, against a wall of a given surface area is simply the integral of the mode distribution.
The pressure, on the other hand, is obtained from simple mathematical relations:

$$p = \frac{2}{3} n \langle E \rangle$$

The pressure is therefore directly connected to the mean value of the energy, therefore also to the mean square velocity and temperature values. With similar considerations it can be deduced that the average kinetic energy is given by:

$$\langle T \rangle = \frac{1}{N} \sum_{i=1}^{N} \frac{1}{2} m v_i^2 = \frac{3}{2} kT$$

Temperature is therefore the measure of the kinetic energy of the molecules of a gas.
For a monatomic gas, the internal energy is:

$$U = \frac{3}{2} nRT$$

The equation of state of perfect gases can be related to the kinetic theory of gases:

$$pV = n_{mol} RT = \frac{1}{3} Nm \langle v^2 \rangle$$

Where N is the total number of particles present in the volume V and m is the molar mass.
If the volumes of the individual particles and the intermolecular forces cannot be neglected, the equation of state transforms into the Van der Waals equation:

$$\left(p + \frac{a n_{mol}^2}{V^2} \right) (V - b n_{mol}) = n_{mol} RT$$

With a and b suitable constants, different for each material.
In this equation there is an isotherm with a horizontal inflection point.
This point is called critical and corresponds to the upper limit of the area of coexistence between liquid and vapor and obviously depends on the material considered.
The values of the state functions at the critical point are given by:

$$T_{cr} = \frac{8a}{27bR}$$

$$p_{cr} = \frac{a}{27b^2}$$

$$V_{cr} = 3bn_{mol}$$

At the critical value, it holds:

$$p_{cr}V_{cr} = \frac{3}{8}RT_{cr}$$

which is different from the general ideal gas equation.
By applying series expansion to pressure as a function of molar volume, the so-called virial expansion is obtained.
The temperature at which the second term of this expansion is zero is called the Boyle temperature.
Twice this value is the inversion temperature.
For a Van der Waals gas the following holds:

$$T_B = \frac{a}{Rb}$$

$$T_i = 2T_B$$

The Boyle temperature is the temperature at which real gases have an ideal behavior for small pressure values.
For a given pressure, a real gas has an inversion temperature.
Above this temperature, an expansion with constant enthalpy causes a heating of the gas while below this temperature there is a cooling.

The equation of state for solids and liquids is instead given by:

$$\frac{V}{V_0} = 1 + \frac{1}{V}\left(\frac{\partial V}{\partial T}\right)_p \Delta T + \frac{1}{V}\left(\frac{\partial V}{\partial p}\right)_T \Delta p$$

Considering a gas as composed of colliding particles, the mean free path between one collision and the next is given by:

$$l = \frac{v_1}{nu\sigma}$$

Where the cross section appears, n is the number of particles, the velocity is that single particle, while u is the quadratic sum of the relative velocities of the colliding particles.
The mean time between two successive collisions is:

$$\tau = \frac{1}{n\sigma v}$$

A gas is said to be Knudsen if the mean free path is much greater than the spatial dimensions in which the gas is confined.
Generally, a gas is Knudsen at low pressure values.
The conductivity of heat in a gas at rest is as follows:

$$\frac{dQ}{dt} = \kappa A\left(\frac{T_2 - T_1}{d}\right)$$

Where d is the generic distance between point 2 and point 1, while k is easily obtained from the kinetic theory of gases and can be related to viscosity:

$$k = \frac{C_{mV} \cdot nl\langle v\rangle}{3N_A} = C_V \eta$$

The temperature profile of the gas at rest with a heat conductivity defined as above is linear:

$$T(z) = T_1 + \frac{z(T_2 - T_1)}{d}$$

On the other hand, a value of k that takes into account nonlinear effects is the one obtained from Eucken's correction:

$$k = \left(1 + \frac{9R}{4c_{mV}}\right) C_V \eta$$

High school exercises

Exercise 1

Gold has a molar mass of 197 g/mol. Take a 2.5 gram sample, determine the number of moles and the number of atoms.

The number of moles is:

$$n_{moli} = \frac{2.50\,g}{197\,\frac{g}{mol}} = 0.013\,mol$$

The number of atoms is given by the product of this quantity and Avogadro's number:

$$n_{atomi} = 0.013 \times 6.02 \cdot 10^{23} = 7.64 \cdot 10^{21}$$

Exercise 2

An ideal gas undergoes this cycle:

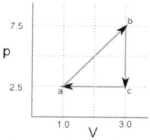

The gas temperature at A is 200 K.

Find the number of moles of the gas, the temperature at B and C, and the net heat supplied to the gas over the entire cycle.

From the graph we deduce:

$$p_a = 2.5\,kN/m^2, \quad V_a = 1.0\,m^3;$$
$$p_b = 7.5\,kN/m^2, \quad V_b = 3.0\,m^3;$$
$$p_c = 2.5\,kN/m^2, \quad V_c = 3.0\,m^3.$$

The ab transformation takes place with an increase in pressure, the bc is isochoric, the ca is isobaric.
At point a we find:

$$n = \frac{pV}{RT} = \frac{2500\,\frac{N}{m^2} \times 1.0\,m^3}{8.31\,\frac{J}{mol\,K} \times 200\,K} = 1.5\,moli$$

Since the number of moles is constant, we have:

$$T_b = \frac{pV}{nR} = \frac{7500\,\frac{N}{m^2} \times 3.0\,m^3}{1.5\,mol \times 8.31\,\frac{J}{mol\,K}} = 1805\,K$$

$$T_c = \frac{2500\,\frac{N}{m^2} \times 3.0\,m^3}{1.5\,mol \times 8.31\,\frac{J}{mol\,K}} = 602\,K$$

In a cyclic transformation the internal energy of the system does not vary and therefore Q=L.
Therefore:

$$Q = \frac{2.0 \times 5.0}{2} = 5.0\,J$$

Exercise 3

Calculate the root mean square velocity of the carbon dioxide molecules at 27°C knowing that the molecular weight is equal to 44 g/mole.

The mass of a molecule is given by:

$$M = \frac{0.044\,g}{6.02 \cdot 10^{23}} = 7.31 \cdot 10^{-26}\,g$$

The root mean square rate is given by:

$$v_{qm} = \sqrt{\frac{3kT}{M}}$$

Where k is Boltzmann's constant:

$$k = \frac{R}{N_A} = 1,38 \cdot 10^{-23}\,\frac{J}{K}$$

Substituting, we get:

$$v_{qm} = \sqrt{\frac{3 \cdot 1.38 \cdot 10^{-23}\,\frac{J}{K} \cdot 300\,K}{7.31 \cdot 10^{-26}\,kg}} = 412\,\frac{m}{s}$$

Exercise 4

Consider air an ideal gas.
Calculate the mean free path of the air molecules assuming as molecular diameter the value of:

$$d = 2,74 \cdot 10^{-10}\,m.$$

The mean free path is given by:

$$\lambda = \frac{1}{4\sqrt{2}\pi r^2 n}$$

Where n is the number of molecules and:

$$4\pi r^2$$

It is the effective section for shocks.
From the ideal gas equation of state:

$$n = \frac{1,013 \cdot 10^5 \, \frac{N}{m^2}}{1,38 \cdot 10^{-23} \, \frac{J}{K} \cdot 293 \, K} = 2,51 \cdot 10^{25} \, \frac{molecole}{m^3}$$

So the mean free path is:

$$\lambda = \frac{1}{4\sqrt{2}\pi \cdot (2,74 \cdot 10^{-10} \, m)^2 \cdot 2,51 \cdot 10^{25} \, \frac{molecole}{m^3}} = 2,99 \cdot 10^{-8} \, m$$

Exercise 5

At what temperature is the root mean square velocity of nitrogen molecules equal to that possessed by hydrogen molecules at 27°C?

From the principle of equipartition we have:

$$\frac{1}{2} m v_{qm}^2 = \frac{3}{2} kT$$

From which we obtain for hydrogen:

$$v_{qm} = \sqrt{\frac{3kT_H}{m_H}}$$

And then:

$$\sqrt{\frac{3kT_H}{m_H}} = \sqrt{\frac{3kT_N}{m_N}}$$

Solving:

$$T_N = T_H \cdot \frac{m_N}{m_H} = 300\,K \cdot 14 = 4200\,K$$

University-level exercises

Exercise 1

In a container with a volume of 20 liters there is half a mole of nitrogen at 27°C.
What is the average translational energy, the average speed of the molecules, the total energy of each single molecule and the internal energy of the gas?

The molecular weight of diatomic nitrogen is 28.
The mean translational energy is:

$$E_{tr} = \frac{3}{2}kT = \frac{3}{2} \cdot 1.38 \cdot 10^{-23}\,J/K \cdot 300\,K = 6.2 \cdot 10^{-21}\,J$$

The average speed of the molecules is:

$$v = \sqrt{\frac{2E_{tr}}{m}}$$

Or:

$$v = \sqrt{\frac{3RT}{PM}} = \sqrt{\frac{3 \cdot 8.31\,J/mole\,K \cdot 300\,K}{28\,g/mole}} = \sqrt{\frac{3 \cdot 8.31\,J/mole\,K \cdot 300\,K}{28 \cdot 10^{-3}\,kg/mole}} = 520\,m/s$$

There are 5 degrees of freedom (3 translational and 2 rotational) so the total energy of each molecule is:

$$E_{tot} = \frac{5}{2}kT = 1.03 \cdot 10^{-20}\,J$$

The total internal energy is the sum over the number of molecules:

$$U = \frac{5}{2}nRT = 3.1\,kJ$$

Exercise 2

0.2 moles of monatomic ideal gas undergo the following transformation:

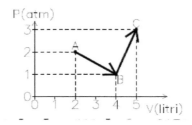

Calculate the change in internal energy due to the change in temperature of the molecules.

The average kinetic energy of a molecule is:

$$E_{cin} = \frac{\nu}{2}kT$$

The total change in internal energy is:

$$\Delta U = \Delta E_{cin} \cdot N_{molecole} = \frac{\nu}{2}\frac{R}{N_A}\Delta T \cdot N_{molecole} = \frac{\nu}{2}R\frac{N_{molecole}}{N_A}\Delta T = \frac{\nu}{2}Rn\Delta T$$

There are 3 degrees of freedom for a monatomic gas:

$$\Delta U = \frac{\nu}{2}nR\Delta T = \frac{3}{2}nR\Delta T = \frac{3}{2}0.02\,moli \cdot 8.31J/mole\,K(915 - 244)\,K = 1670\,J$$

11

ELECTROMAGNETISM

Background

Electric and magnetic phenomena were unknown to primitive man and the first human civilizations that populated la MesopotamiaEgypt andla Persia. The first studies on these phenomena date back to ancient Greece by the philosopher Thales towards the sixth century BC.

The name electricity that we all use derives from the properties discovered by Thales of some fossil resins, such as amber, which became electrified by friction, in fact in ancient Greek amber was called "electron".

Some hypothesize that the classic behavior of magnetite in attracting iron filings was previously known in China, where the first rudimentary compass was also built, but this assumption is not supported by archaeological and historical discoveries.

Centuries later, in ancient Rome in the first century after Christ, both Pliny the Elder and Seneca described the properties of amber and investigated the types of lightning.

These ancient studies were followed by over a millennium of silence on these phenomena with the exception of the English monk Venerable Bede who described, towards the eighth century, properties similar to those of amber in other materials and the study of Pietro Peregrino in which introduced for the first time the terminology of North Pole and South Pole and in which the attractive and repulsive properties of magnets were studied.

The first scientific studies on electric and magnetic phenomena began much later, after the scientific and cultural revolution of the Renaissance and Copernicus and after the definition and subsequent affirmation of the modern scientific method introduced by Galileo Galilei.

Among other things, the same scientists who undertook the task of describing these phenomena, including Galileo and Newton, made gross errors by attributing the electrical and magnetic properties of certain materials to strange effluvia or air movement.

In the eighteenth century a considerable quantity of electric and magnetic phenomena were discovered, even if these inventions almost always remained disconnected from each other and without an apparent logical thread that united them.

It began to be understood that both electricity and magnetism could be repulsive or attractive in the sense that, by changing certain conditions, one could witness a force that moved the electrified objects away or closer. This created further misunderstandings, as the majority of scientists began to think that there were two electric fluids, one "glassy" i.e. with a positive charge and one "resinous" i.e. with a negative charge.

The names vitreous and resinous referred to the different behavior that glass and resins, including amber, had when electrified by friction.

In the mid-eighteenth century, the American Benjamin Franklin, based on some correct observations of electrical phenomena, designed the first lightning rod and the first systems went into operation only six years after the first positive experiment.

Coulomb's law

The first scientific studies on electrical phenomena were carried out at the end of the eighteenth century by Cavendish and Coulomb.

In particular, the second scientist formulated, in 1785, the law that bears his name and which expresses the electric force generated by two electric charges:

$$F = k \frac{q_1 q_2}{d^2}$$

Where d is the distance between the electric charges and k is Coulomb's constant.

This force can be attractive or repulsive based on the discordant or concordant sign of the electric charges.

This equation is very similar to that of Newton's universal gravitation, with the only difference that the latter is only attractive.

Therefore, in Coulomb's studies the electric force is a force that acts instantaneously at a distance.

Coulomb's constant is related to the dielectric constant like this:

$$k = \frac{1}{4\pi\varepsilon}$$

Volta's laws and Ampére's law

At the beginning of the nineteenth century Volta built the first electrochemical battery, demonstrating some characteristics of electricity and enunciating the famous laws that bear his name, especially as regards the nature of metals, the concept of electric conductor and electric potential .

Unlike Galvani and his experiments with frogs, Volta understood that the electricity did not depend on the animal but on the metallic conductor, in particular on the pair of metallic conductors used. To demonstrate this, he conducted experiments with zinc and silver, which he had found to be the most effective pair of dissimilar metals, by constructing two different pieces of these metals placed in beakers filled with brine.

Volta had built the first electric cell in which zinc and silver represented the positive and negative electrodes and the brine represented the saline solution that is needed to trigger the electricity generating mechanism.

Volta gave unequivocal proof of the nature of electric current seen as a flow of electrically "charged" particles flowing between one material and another.

Hand in hand, Hans Christian Oersted, a well-known Danish academic professor with considerable studies on electricity behind him, in 1820 discovered, almost by accident, the influence that an electric current could have on the needle of a magnetic compass.

While he was preparing material for a university lesson, a compass ended up near an electric wire in which electricity was flowing and Oersted saw that the compass needle suddenly moved.

Repeating the experiment, he discovered that, in the absence of electricity, the magnetic needle correctly marked the classic north-south direction, while in the presence of electric current in the wire, the needle deviated completely from this direction until it was even perpendicular, if the current was of a fairly high intensity.

For the first time ever, Oersted had demonstrated that electricity affected the magnetic behavior of materials and established a first relationship between electrical phenomena and magnetic phenomena.

In 1826 Ampére enunciated a law linking electric and magnetic phenomena, after having generalized the observations made by Oersted concerning the creation of magnetic fields by parts of electric currents.

This law states that the integral of the magnetic field on a closed line is given by the sum of the chained electric currents multiplied by the magnetic permeability.

$$\oint_{\partial S} \vec{B} \cdot d\vec{r} = \mu I_E$$

This integral is not zero and this implies that the magnetic field is not conservative.

By applying the well-known Gauss flux theorem to the electric and magnetic fields, the following Gauss laws were obtained:

$$\nabla \cdot D = \rho_E$$
$$\nabla \cdot B = 0$$

Faraday and Neumann

Barely a year after Ampère's experiment, Faraday understood that to discover new scientific realities it was necessary to overturn the meaning of the experiments carried out up to that moment; therefore he decided to place a metal wire inside a cavity in which there were magnets of the opposite sign and he noticed that the wire moved when it was traversed by electric current.

For the first time, Faraday realized that magnetism could also affect electrical behavior.

He devised an experiment in which he wound two coils of insulated wire around a large steel core and discovered, to his surprise, that a current on the first coil induced, thanks to the magnetic action of the steel core, a current on the second coil as well. ; magnetism had actually changed the electrical behavior of materials.

Faraday used these forces to physically move objects in a rotational direction.

Thus it was that he created the first transformer and the first electric motor in history.

In 1831, Faraday deduced the influence that magnetic fields had on the creation of currents (called induced).

The obtained law was called magnetic induction.

The potential difference created, called electromotive force, was thus linked to the magnetic flux:

$$\Delta V = -\frac{\partial}{\partial t} \int_{\Sigma(t)} \vec{B}(\vec{r},t) \cdot d\vec{A}$$

This law was then generalized by Neumann and Lenz in the following years.

Maxwell's equations

However, the experimental and theoretical system built during the first half of the 19th century lacked a single coding and a unitary way of understanding electric and magnetic phenomena.

The real turning point came in 1864 with the publication of the well-known Maxwell's equations, whose elegance and completeness in the treatment are still today a source of admiration and a great test bench for the enormous theoretical and practical consequences.

Very briefly, these equations in differential form are the following :

$$\nabla \cdot D = \rho_E$$

$$\nabla \times E + \frac{\partial B}{\partial t} = 0$$

$$\nabla \cdot B = 0$$

$$\nabla \times H - \frac{\partial D}{\partial t} = J_E$$

where E is the electric field, D the electric induction, H the magnetic field, B the magnetic induction, ρ_E the electric charge density and J_E the current density, all vector quantities.

The constitutive relations which relate the electric and magnetic fields with the respective induction vectors are, in linear, homogeneous, stationary and isotropic media (just to simplify and not to put matrix and tensor notations), the following:

$$D = \varepsilon_0 \varepsilon_r E$$

$$B = \mu_0 \mu_r H$$

the above constants are the dielectric constants and the magnetic permeabilities (absolute with the subscript zero, relative with the subscript r).

The absolute constants are those relevant to the vacuum and are thus related to each other

$$\varepsilon_0 \mu_0 = \frac{1}{c^2}$$

and c is the speed of light.

Maxwell's equations reveal multiple facets, the list of which can only be reported by dedicating an entire book to it.
Here we try to give an extreme summary of the main issues.

1) First of all, the four equations represent a generalization of everything discovered up to then in electric and magnetic terms.
The first equation given is nothing but the electric Gauss law (and the third, the magnetic Gauss law), while the second is Faraday's law and the last is an extension of Ampére's law made by the same Maxwell.

2) From the third equation, in particular, the non-existence of isolated magnetic monopoles is determined which, stated in another way, means admitting lines of force of the magnetic field which are closed.

3) From the first equation it can be seen that the lines of force of the electric field are, by definition, open, i.e. with a starting point and an arrival point determined by the electric charges of opposite sign which generate the field itself.

4) Combining instead the first, the second and the fourth equation we obtain the already known conservation of the electric charge or continuity equation:

$$\nabla \cdot J_E + \frac{\partial \rho_E}{\partial t} = 0$$

5) In an electromagnetic field, the force to which a charge is subjected is given by the expression of the Lorentz force, which is an overcoming and generalization of the Coulomb force:

$$F = q(E + \mu v \times H)$$

where v is the speed at which charge q is moving.

6) A further point of discussion arises from the fact that in the second and fourth equations the electric and magnetic fields are mutually bound to each other.

This evidence put an end to almost 50 years of misunderstandings between electrical and magnetic phenomena, simply by unifying them.
Since 1864 inthen it was clear that there was a single field with a single force that described a single concept and since then all this took the adjective electromagnetic.
It is true that at times the electric or magnetic effects prevail, but in reality the two entities are never present only individually.

7) Another aspect, closely linked to the previous one, is that of duality.
It is clearly seen that the equations are somehow specular with respect to the electric and magnetic fields, almost managing to swap their roles.
Indeed, under a suitable rewriting of Maxwell's equations, this is possible in a natural and automatic way.

8) Another consequence is the overcoming of Coulomb's theory of action at a distance with a local point-to-point interaction between electric phenomena and magnetic phenomena, so there was an extension of the notions much more than the Lorentz force can make believe.

The electromagnetic waves

But of all the most striking effects, there was one that was impressive because it foresaw something not yet experimentally proven and because it radically modifies the very idea of some basic concepts.
To the fundamental question of how electric and magnetic phenomena were transmitted, Maxwell's equations gave a disconcerting answer: in the form of waves.
This assumption derives essentially from the solutions of Maxwell's equations.

Said A a generic vector potential and ϕ a generic scalar potential, the first two Maxwell equations can be rewritten as follows

$$\nabla \times A = B$$

$$\nabla \phi + \frac{\partial A}{\partial t} = -E$$

and substituting in the other two Maxwell's equations the equations for the potentials are obtained.

$$\nabla^2 \phi - \frac{1}{c^2} \frac{\partial^2 \phi}{\partial t^2} = -\frac{\rho_E}{\varepsilon}$$

$$\nabla^2 A - \frac{1}{c^2} \frac{\partial^2 A}{\partial t^2} = -\mu J_E$$

whose form is completely identical to the D'Alembert wave equation, where c is the propagation speed of these waves.

These waves, solutions of the equations, (henceforth called electromagnetic waves) have a particular property: they do not require a means of transmission and can also propagate in a vacuum with a speed which is that of light, as can be clearly seen from the above equations exposed.

The empirical evidence of what has just been stated was found in 1885 with Hertz's experiments in which, not only did what Maxwell predicted years earlier was confirmed, but the fundamental foundations for telecommunications as we know them today were laid.

From that moment it was evident that the transmission of information by means of electromagnetic waves could be detached from a physical support such as electric cables using the air itself in their place.

We owe it to all of this that a few decades later radio, television and much later satellite communications and cellular telephony were invented.

Hertz's experiments

How can this theoretical system be verified experimentally?

Maxwell's equations, in fact, predicted effects not yet known on a practical level such as for example the propagation of the electromagnetic field in the form of waves that did not require a means of transport.

It must be said that Maxwell's great intuition was not immediately followed by a horde of experiments as the physicists of the time had first to become familiar with that new formalism.

It took twenty-one years for definitive proof.

In 1885, the German physicist Hertz devised the following experiment to test whether or not Maxwell's equations were true.

Hertz first had to devise a potential source of electromagnetic waves.

Without it, it would have been impossible to think of testing Maxwell's theory.

Taking advantage of what Faraday studied about the currents induced by a magnetic field, he took a coil and created a high potential difference between two conductors placed at a certain distance.

322

By increasing this potential difference, the dielectric limit would have been reached such that an electric discharge would have been triggered between the two conductors.

According to Maxwell's theory, this discharge, making rapid oscillations, should have generated electromagnetic waves.

At that point Hertz had to devise a detector of such possible waves.

He took a copper wire and wound it into an open loop.

He fixed one side of the coil to a metal conductor (a bronze sphere) while the other side left it free and placed it a short distance from the bronze sphere.

If an electromagnetic wave had been generated, as Maxwell's equations predicted, then it would have induced an electromagnetic field in the coil causing a spark to strike between the bronze sphere and the free side of the coil.

Indeed, the experiment worked and Hertz managed to perfect it by studying the typical characteristics of waves, such as polarization and reflection.

Maxwell was therefore right: the electromagnetic field propagated in the form of waves that did not require a transmission medium.

From Hertz's experiment onward it became apparent that electric and magnetic fields could use air to be transmitted and that conductors (such as copper wire) were not entirely necessary.

We owe to these experiences if, only a few decades later, it was possible to use electromagnetic waves to replace the telegraph.

Radio telegraphy was the forerunner of radio or radio broadcasting.

Characters like Marconi and Tesla owe much to the Hertz experiment and Maxwell's equations.

From radio we soon moved on to television and from here to contemporary telecommunications such as satellite and cellular ones.

We can state that, without the Hertz experiment, today's technology would not be the same.

We note that the first radio broadcasts began only about twenty years after the first Hertz experiment.

Maxwell's equations had taken twenty years to verify, and after a similar period, there was already a world-class industry with thousands of local radio stations and ever-widening applications, from civilian to military.

Optical phenomena

A final observation, perhaps of a more theoretical than practical nature at least in the nineteenth century, concerned the convergence of another branch of physics within the theory of electromagnetism.

All experiments and studies about optics and light rays became a part of electromagnetism in the sense that what we call visible light is a part of the electromagnetic spectrum, between certain wavelengths that our eyes can detect.

Below the minimum visibility threshold (corresponding to the red colour) there are electromagnetic waves such as infrared and radio waves (those used for telecommunications, to be clear), above the maximum visibility threshold (corresponding to the violet colour) there are ultraviolet and cosmic rays.

Colors are therefore only manifestations of the properties of electromagnetic waves in terms of reflected and absorbed wavelengths.

The light therefore follows the same equations of Maxwell, like any other electromagnetic wave.

A decisive consequence of this convergence between electromagnetism and optics concerns astronomy and life itself: the starry sky exists precisely by virtue of the property of electromagnetic waves to be transmitted in a vacuum and the heat of the Sun can be transmitted at a distance precisely because the waves electromagnetic waves, and therefore light, can travel in a vacuum.

General properties of waves

Maxwell's equations confront us with the fundamental problem of mathematically defining wave phenomena.

In the history of modern physics this need has persisted over the centuries and has had various approaches, first starting from mechanics, later arriving at optics and finally arriving at studies on electromagnetism.

A wave is a periodic perturbation of a given physical entity that transports energy without transporting matter.

Periodicity characterizes every wave phenomenon and we have clearly seen it by introducing wavelength and frequency, which are nothing but the expression of periodicity in spatial and temporal terms, respectively.

In particular, we can define the wavelength as the periodic spatial distance between two particular points, for example the maxima of the wave amplitude (called peaks) or the points where this amplitude is zero (called nodes).

The amplitude instead takes into account the power of the wave.

The fundamental relationship between wavelength and frequency in a material medium is given by:

$$\lambda f = \frac{\lambda_0}{n} f = v = \frac{v_0}{n}$$

Where v is the speed of the wave, zero subscripts account for physical quantities in vacuum, while n is the refractive index.
As can be seen, the speed of a wave depends on the material medium in which it propagates and this dependence is transferred entirely to the wavelength, while the frequency remains constant.
We can define two further quantities related to the previous ones: the wave number and the pulsation.

$$k = \frac{2\pi}{\lambda}$$

$$\omega = 2\pi f$$

The general equation of a wave was explained in 1747 by D'Alembert:

$$\nabla^2 u - \frac{1}{v^2} \frac{\partial^2 u}{\partial t^2} = 0$$

Being a partial differential equation, it will have different solutions based on the spatial boundary conditions and the temporal initial conditions.
If the wave number and the velocity are parallel to the solution of this equation, then we speak of longitudinal waves and the wave oscillates in the same direction in which it propagates.
If, on the other hand, they are perpendicular, we have the case of transversal waves.
Two different concepts of speed can be defined for a wave.
The phase velocity is given by the ratio between pulsation and wave number, the group velocity by their derivative.

$$v_{ph} = \frac{\omega}{k}$$

$$v_g = \frac{d\omega}{dk} = v_{ph}\left(1 - \frac{k}{n}\frac{dn}{dk}\right)$$

These speeds coincide only if the medium is not dispersive.
Conversely, the group velocity can be less or greater than the phase velocity.

The group speed is the speed with which the energy contribution linked to the wave is transferred and, therefore, characterizes much better the description of a wave phenomenon.

The general solution of the D'Alembert equation in one dimension is given by:

$$u(x,t) = \int_{-\infty}^{+\infty} \left(a(k)e^{i(kx-\omega_1(k)t)} + b(k)e^{i(kx-\omega_2(k)t)} \right) dk$$

This Fourier integral depends both on the boundary conditions and on the dispersion relations.
In Cartesian coordinates, the solutions of D'Alembert's equations are called plane waves.
In one dimension this solution is given by a stationary wave, ie by a wave in which the peaks and nodes correspond to fixed and defined spatial points.
A plane wave is defined by:

$$u(\vec{x},t) = |\vec{u}|\cos\left(\vec{k} \cdot \vec{x} \pm \omega t + \varphi\right)$$

Where φ is the phase, while the sign $\pm$ takes into account the progressive wave and the regressive one (ie its reflection when it arrives at the boundary).
If the boundary condition dictates that the plane wave is zero at the boundary, then the reflected wave will undergo a 90° phase change.
If, on the other hand, the imposition is given by the cancellation of the derivative around the edge, there is no phase change.
An observer in motion will notice a change in frequency of the wave, called Doppler effect, particularly important in acoustics, optics and electromagnetism:

$$\Delta f = \frac{v_f - v_{obs}}{v_f}$$

In spherical coordinates, the wave equation for the radial is given by:

$$\frac{1}{v^2}\frac{\partial^2(ru)}{\partial t^2} - \frac{\partial^2(ru)}{\partial r^2} = 0$$

With a general solution like this:

$$u(r,t) = C_1 \frac{f(r-vt)}{r} + C_2 \frac{g(r+vt)}{r}$$

In cylindrical coordinates, the wave equation for the radial is the Bessel equation:

$$\frac{1}{v^2} \frac{\partial^2 u}{\partial t^2} - \frac{1}{r} \frac{\partial}{\partial r}\left(r \frac{\partial u}{\partial r} \right) = 0$$

The solutions are the Hankel functions. For sufficiently large values of r, the following holds:

$$u(r,t) = \frac{|\bar{u}|}{\sqrt{r}} \cos\big(k(r \pm vt)\big)$$

For all these equations, the analytical solutions can be calculated only in special cases of symmetry and fairly regular boundary conditions.
In all other cases, it is extremely difficult to determine the analytic solutions also because the general Fourier integral cannot be solved in absolute terms.
To overcome this, it is necessary to resort to numerical solutions through suitable methods of numerical calculation and discretization of physical quantities.
The wave equation describes wave phenomena regardless of their physical nature. Therefore it is possible to study the mechanical waves generated by a vibrating string or by an oscillating membrane or the acoustic waves or the waves of an earthquake or the pressure waves in a gas or the electromagnetic waves.
In addition, propagation mechanisms such as shock waves (for example ultrasonic waves) or solitonic waves (typically those of a tsunami) can also be explained.
Each type of wave just described has typical characteristics which uniquely identify it and which are defined only by probing each individual discipline it belongs to (acoustics, seismics, thermodynamic kinematics, electromagnetism and so on).

In particular, for electromagnetic waves, Maxwell's equations are valid.
There can be three different cases for the boundary conditions: transverse magnetic modes, transverse electric modes or both transverse.

In the first two cases, eigenvalue equations and a cut-off frequency are obtained below which it is not possible to confine the wave in a guided path.

With a similar procedure it is possible to obtain the frequencies of an electromagnetic wave in a resonant cavity and the possible number of oscillation modes.

For a cube-shaped cavity, the frequency is given by:

$$f = \frac{v}{2a}\sqrt{n_x^2 + n_y^2 + n_z^2} = \frac{\sqrt{3}vn}{2a}$$

Where the last equality exists if the medium has a constant and homogeneous refractive index.

In that case, the number of oscillation modes is:

$$N_T = 2\frac{4\pi a^3 f^3}{3v^3}$$

Where the 2 is due to the two states of polarization of the electromagnetic waves.

Tensor notation of the electromagnetic field

With tensor notation, the electromagnetic field can be represented with an antisymmetric double tensor, of the second order and covariant, having zero trace:

$$F^{\mu\nu} = \partial^\mu A^\nu - \partial^\nu A^\mu$$

where A is the generic vector potential previously exposed.

This tensor can also be replaced by its electromagnetic dual $G^{\mu\nu}$ which takes into account the magnetic components.

Furthermore, we have the following remarkable property:

$$\det(F) = \left(\frac{B \cdot E}{c}\right)^2$$

The new wave equation in tensor form is as follows:

$$(\nabla^2 - \frac{1}{c^2}\frac{\partial^2}{\partial t^2})A^\mu = -\mu J_E^\mu$$

While Maxwell's equations can be reduced to two as follows:

$$\partial_\nu F^{\mu\nu} = 0$$

or in an entirely equivalent way

$$\partial_\mu G^{\mu\nu} = 0$$

$$\partial_\mu F^{\mu\nu} = \frac{4\pi}{c}j^\nu$$

The first equation is the synthesis of the magnetic Gauss law and the Faraday law, while the second equation is the union of the electric Gauss law and the Ampére-Maxwell law.

Final note on the nature of light

After investigating geometric optics and electromagnetism, what can we say about light?
Is it a wave or a particle?
From what is stated in this manual, it would seem, beyond a shadow of a doubt, that light is a wave.
It even satisfies the wave equation and propagates like a wave following Maxwell's equations!
We are still missing a piece to determine the real nature of light.
The piece derives from the implications of the Maxwell equations.
Thanks to them, the logical system of classical physics was shattered and, about forty years after the publication of those equations, exactly in 1905 someone (Albert Einstein!) explained the photoelectric effect by introducing corpuscles, called photons.
Thus, dualism reappeared.
It took another twenty years to solve the mystery.
First in 1924, with the studies of De Broglie, it was clear that there was a correlation between material quantities and wave nature.

But it was in 1926, with the publication of Schrodinger's equation, that the mystery was definitively understood.
In fact, Schrodinger's equation unifies the concepts of wave and matter in the so-called "material waves" therefore light is both wave and particle.
We will see soon…

High school exercises

Exercise 1

A given battery has a potential of 12 V and can flow a charge of 84 Ah. Find the corresponding coulomb charge and the energy used to flow the charge.

From the definition of current intensity:

$$i = \frac{\Delta q}{\Delta t} \qquad \Delta q = i \Delta t$$

And then:

$$84 \frac{C \cdot h}{s} \times \frac{3600 \, s}{h} = 302400 \, C$$

The energy used is equal to:

$$\Delta U = \Delta V \cdot \Delta q = 12 \, V \times 302400 \, C = 3.63 \cdot 10^6 \, J$$

Exercise 2

Two parallel conducting plates are placed 12 cm apart.
Equal and opposite charges are present on their surfaces.
An electron placed midway is subjected to a force of

$$3.9 \cdot 10^{-15} \, N.$$

Find the electric field at the position of the electron and the potential difference between the two plates.

From the definition of electric field:

$$E = \frac{F}{e} = \frac{3.9 \cdot 10^{-15}\,N}{1.602 \cdot 10^{-19}\,C} = 2.4 \cdot 10^4\,\frac{N}{C}$$

The potential difference will be:

$$\Delta V = Er = 2.43 \cdot 10^4\,\frac{N}{C} \times 1.2 \cdot 10^{-1}\,m = 2.9 \cdot 10^3\,V$$

Exercise 3

Find the charge and charge density on the surface of a spherical container of radius 15cm whose potential is 200V.

The amount of charge on the surface is:

$$q = \frac{Vr}{k_0} = \frac{200\,V \times 0.15\,m}{8.99 \cdot 10^9\,\frac{Nm^2}{C^2}} = 3.34 \cdot 10^{-9}\,C$$

The charge density is obtained by dividing this quantity by the surface area of the sphere, since the charge is uniformly distributed:

$$\sigma = \frac{q}{A} = \frac{2.34 \cdot 10^{-9}\,C}{4\pi \times 0.15^2\,m^2} = 1.18 \cdot 10^{-8}\,\frac{C}{m^2}$$

Exercise 4

Given two electric charges:

$$q_1 = +3.0 \cdot 10^{-6}\,C$$
$$q_2 = -4.0 \cdot 10^{-6}\,C,$$

Find the work that needs to be done to place these charges in:

$$x = 3.5\,cm,\ y = +0.50\,cm;$$
$$x = -2.0\,cm,\ y = +1.5\,cm$$

The distance between the two charges is:

$$r = \sqrt{(3.5 + 2.0)^2 + (0.50 - 1.50)^2} = 5.6\,cm$$

The work is given by:

$$L = \frac{1}{4\pi\varepsilon_0}\frac{q_1 q_2}{r} = 8.99\cdot 10^9\,\frac{Nm^2}{C^2}\times\frac{(-12\cdot 10^{-12})\,C^2}{0.056\,m} = -1.9\,J$$

Exercise 5

Two electrons are held fixed 2 cm apart. A third electron, coming from infinity, stops midway.
Calculate its initial speed.

The potential at the midpoint of the segment is given by:

$$V = \frac{2e}{4\pi\varepsilon_0 r} = \frac{8.99\cdot 10^9\,\frac{N}{m^2 C^2}\times 2\times 1.60\cdot 10^{-19}\,C}{1.0\cdot 10^{-2}\,m} = 2.9\cdot 10^{-7}\,V$$

The initial velocity is zeroed due to the repulsive force acting on the third electron.
According to the principle of conservation of energy, the work required is equal to the kinetic energy, i.e.:

$$-Ve = \frac{1}{2}m\left(v_f^2 - v_0^2\right) = -\frac{1}{2}mv_0^2$$

From which:

$$v_0 = \sqrt{\frac{2eV}{m}}$$

In our case we have:

$$v_0 = \sqrt{\frac{2eV}{m}} = \sqrt{\frac{2 \times 1.60 \cdot 10^{-19}\,C \times 2.9 \cdot 10^{-7}\,V}{9.11 \cdot 10^{-31}\,kg}} = 319\,\frac{m}{s}$$

Exercise 6

Calculate the strength of the gravitational force and the electric force in the hydrogen atom.

The hydrogen atom consists of a proton and an electron at an average distance equal to the average atomic radius:

$$r = 0.53 \cdot 10^{-10}\,m.$$

The electric force is given by:

$$F_e = \frac{1}{4\pi\varepsilon_0}\frac{qq_0}{r^2} = \frac{9 \cdot 10^9 \cdot \left(1.60 \cdot 10^{-19}\right)^2}{\left(0.53 \cdot 10^{-10}\right)^2} = 8.20 \cdot 10^{-8}\,N$$

The gravitational force is instead:

$$F_g = G\frac{m_1 m_2}{r^2} = \frac{6.67 \cdot 10^{-11} \cdot 9.11 \cdot 10^{-31} \cdot 1.67 \cdot 10^{-27}}{\left(0.53 \cdot 10^{-10}\right)^2} = 3.61 \cdot 10^{-47}\,N$$

Much smaller than the first.

Exercise 7

A conducting sphere of mass 2 grams with a charge equal to 0.02 millionth of a Coulomb is suspended along a string of length l.

A second conducting sphere with a charge equal to 0.5 millionth of a Coulomb is brought near the first.
Calculate the value of the angle that the string makes with the vertical when the distance between the two spheres is 5 cm.

At equilibrium, the resultant of the weight force and the electric force on the first sphere is directed in the direction of the thread and is balanced by the tension in the thread.
By the theorems of trigonometry we have:

$$\tan \theta = \frac{F_e}{F_g} = \frac{qq_0}{4\pi\varepsilon_0 r^2 mg} = \frac{9 \cdot 10^9 \cdot 5 \cdot 10^{-7} \cdot 2 \cdot 10^{-8}}{25 \cdot 10^{-4} \cdot 2 \cdot 10^{-3} \cdot 9.8} = 0.1837$$

Therefore:

$$\theta = \arctan 0.1837 \cong 10.41°$$

Exercise 8

Find the distance that separates a point charge of 26 millionths of a Coulomb from one of -47 millionths of a Coulomb so that the attractive force is 5.7 N.

Just apply Coulomb's law:

$$d = \sqrt{\frac{kq_1 q_2}{F}} = \sqrt{\frac{8.99 \cdot 10^9 \times 26.0 \cdot 10^{-6} \times (47.0 \cdot 10^{-6})}{5.70}} = 1.39\,m$$

Exercise 9

Two charges of 20 millionths of a Coulomb are held stationary 1.5 meters apart.
Calculate the force on one of them.
A third identical charge is approached and placed as in the figure:

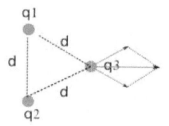

Calculate the force on the first charge.

Initially the force is:

$$F = \frac{1}{4\pi\varepsilon_0} \frac{q_1 q_2}{d^2} = 8.99 \cdot 10^9 \frac{(20.0 \cdot 10^{-6})^2}{1.50^2} = 1.60\,N$$

The third charge is placed to form an equilateral triangle.
Carrying out the vector addition of the forces, we have:

$$F = 2 \times 1.60 \times \frac{\sqrt{3}}{2} = 2.77\,N$$

Exercise 10

Two charges Q are attached to the two opposite corners of a square.
Two charges q are placed on the other vertices.
If the electric force is on Q, evaluate the relationship between q and Q.

If the net force on Q is zero, q must be of opposite sign and such that the forces exerted on Q are the horizontal and vertical components of the repelling force.
The repulsive force between the two charges Q is:

$$F = k \frac{Q^2}{2l^2}$$

The attractive force between q and Q is:

$$F = \sqrt{2\left(\frac{kqQ}{l^2}\right)} = k\frac{qQ}{l^2}\sqrt{2}$$

These forces must therefore be opposite:

$$Q = -2\sqrt{2}q$$

Exercise 11

The electrostatic force between two identical ions separated by 0.5 nm is 3.7 nN.
Find the charge of each ion and the number of missing electrons.

Applying Coulomb's law:

$$q = \sqrt{\frac{Fr^2}{k_0}} = \sqrt{\frac{3.7 \cdot 10^{-9}\,N \times (5.0 \cdot 10^{-10})\,m^2}{8.99 \cdot 10^9}} = 3.21 \cdot 10^{-19}\,C$$

The number of missing electrons is simply given by dividing by the electron charge:

$$n = \frac{q}{e} = \frac{3.21 \cdot 10^{-19}\,C}{1.602 \cdot 10^{-19}\,C} = 2$$

Exercise 12

Two point charges 50 cm apart have intensity:

$$q_1 = 2.0 \cdot 10^{-8}\,C \text{ e } q_2 = -4.0q_1$$

Find the point on the axis joining them such that the electric field is zero.

336

We take the origin in the first charge and denote by x the coordinate along the axis joining the two charges.
The electric field will be zero when:

$$E_{tot} = k_0 \frac{2.0 \cdot 10^{-8}}{|x|^2} - k_0 \frac{8.0 \cdot 10^{-8}}{|x - 0.5|^2} = 0$$

From which:

$$3x^2 + x - 0.25 = 0$$

Therefore:

$$x = \frac{-1 \pm \sqrt{1 + 3}}{6} = \begin{array}{l} \frac{1}{6} = 0.17\,cm \\ -0.5\,cm \end{array}$$

For how we have constructed the coordinates, the solution is the negative one.

Exercise 13

An electron moves between two parallel charged plates with velocity components given by:

$$v_x = 1.5 \cdot 10^5\,m/s \quad e \quad v_y = 3.0 \cdot 10^3\,m/s.$$

If the electric field is:

$$\mathbf{E} = (120\,N/C)\,\mathbf{j},$$

Find the acceleration of the electron and its velocity when the x coordinate is changed by 2 cm.

The electric field is directed along the y-axis, the electric force is given by:

$$F = Ee = 120 \frac{N}{C} \times 1.602 \cdot 10^{-19}\,C = 1.92 \cdot 10^{-17}\,N$$

The acceleration is directed downwards:

$$a = \frac{F}{m} = \frac{1.92 \cdot 10^{-17}\,N}{9.11 \cdot 10^{-31}\,kg} = \left(-2.11 \cdot 10^{13}\,\frac{m}{s^2}\right) \mathbf{j}$$

The electron will travel 2 cm in the time:

$$t = \frac{s}{v_x} = \frac{0.020\,m}{1.5 \cdot 10^5\,\frac{m}{s}} = 1.33 \cdot 10^{-7}\,s$$

And the change in velocity along the y axis will be:

$$v_{yf} = v_{yi} + at = 3.0 \cdot 10^5\,\frac{m}{s} - 2.11 \cdot 10^{13}\,\frac{m}{s^2} \times 1.33 \cdot 10^{-7}\,s = -2.51 \cdot 10^6\,\frac{m}{s}$$

Exercise 14

A square surface with side 3.2 mm is immersed in a uniform electric field of 1'800 N/C.
The field lines form an angle of 35° with the outgoing normal.
Calculate the flow through the surface.

The flux of the electric field is the dot product of the field and the area. Considering the angles we have:

$$\Phi = EA\cos\theta$$

I our case:

$$\Phi = 1800\,\frac{N}{C} \times \left(3.2 \cdot 10^{-3}\right)^2 m^2 \times \cos 35° = 1.51 \cdot 10^{-2}\,\frac{N m^2}{C}$$

Exercise 15

A uniformly charged conducting sphere of radius 1.2 meters has a surface charge density:

$$8.1\,\mu C/m^2$$

Find the charge on the sphere and the total electrical flux out.

The area of the sphere is:

$$S = 4\pi r^2 = 4\pi \times 1.2^2 = 18.1 \, m^2$$

The amount of charge is:

$$Q = \sigma S = 8.1 \cdot 10^{-6} \, \frac{C}{m^2} \times 18.1 \, m^2 = 1.5 \cdot 10^{-4} \, C$$

The flow will be:

$$\Phi = \frac{Q}{\varepsilon_0} = \frac{1.5 \cdot 10^{-4} \, C}{8.85 \cdot 10^{-12} \, \frac{C^2}{Nm^2}} = 1.7 \cdot 10^7 \, \frac{Nm^2}{C}$$

Exercise 16

A cylinder 42 cm long and 12 cm in diameter has an electric field of 230,000 N/C.
Find the total charge on the cylinder.
If we reduce the cylinder to 28 cm long and 8 cm in diameter, what is the charge to maintain the same electric field?

The electric field generated by a cylinder is:

$$E = \frac{\lambda}{2\pi\varepsilon_0 r}$$

And so the linear charge is:

$$\lambda = 2\pi\varepsilon_0 rE = 2\pi \times 8.85 \cdot 10^{-12} \, \frac{C^2}{Nm^2} \times 0.06 \, m \times 2.3 \cdot 10^5 \, \frac{N}{C} = 7.67 \cdot 10^{-7} \, \frac{C}{m}$$

The charge will therefore be:

$$Q_1 = \lambda h = 7.67 \cdot 10^{-7} \frac{C}{m} \times 0.42\,m = 3.22 \cdot 10^{-7}\,C$$

After the change of dimensions, imposing the equality of the electric fields as required:

$$\frac{Q_1}{2\pi\varepsilon_0 r_1 h_1} = \frac{Q_2}{2\pi\varepsilon_0 r_2 h_2}$$

The new charge is found:

$$Q_2 = Q_1 \frac{r_2 h_2}{r_1 h_1} = 3.22 \cdot 10^{-7}\,C \times \frac{0.04 \times 0.14}{0.06 \times 0.21} = 1.43 \cdot 10^{-7}\,C$$

Exercise 17

A charge is uniformly distributed in an infinitely long cylinder of radius R. Write the expression for the electric field inside and outside the cylinder.

Inside the cylinder we will have r<R.
The electric field is, applying Gauss's law to the cylinder:

$$E = \frac{q}{2\pi\varepsilon_0 r L}$$

Knowing that:

$$q = \rho V$$

EV is the volume we have:

$$E = \frac{1}{2\pi\varepsilon_0 r L} \cdot \pi r^2 L \rho = \frac{\rho r}{2\varepsilon_0}$$

Outside the cylinder, we have r>R.

We will have to consider all the charge inside the cylinder equal to:

$$q = \rho \pi R^2 L.$$

Gauss' law in this case gives:

$$E = \frac{q}{2\pi\varepsilon_0 rL} \cdot \rho \pi R^2 L = \frac{\rho R^2}{2\varepsilon_0 r}$$

It can be deduced that inside the cylinder the field grows proportionally to r, while outside it decreases in an inversely proportional way.

Exercise 18

A spherical shell has a radius of 25 cm and a charge of 0.2 millionth of a Coulomb.
Find the electric field for a point inside the shell, on its surface 3 meters from the center.

There is no charge inside the shell and therefore the electric field will be zero.
On the surface, applying Gauss's law to a spherical surface we have:

$$E = \frac{q}{4\pi\varepsilon_0 r^2} = \frac{2.0 \cdot 10^{-7}\,C}{4\pi \cdot 8.85 \cdot 10^{-12} \cdot 0.25^2} = 2.9 \cdot 10^4\,\frac{N}{C}$$

While at 3 meters from the shell this value is:

$$E = \frac{q}{4\pi\varepsilon_0 r^2} = \frac{2.0 \cdot 10^{-7}\,C}{4\pi \cdot 8.85 \cdot 10^{-12} \cdot 9} = 200\,\frac{N}{C}$$

Exercise 19

Given the following electrical circuit:

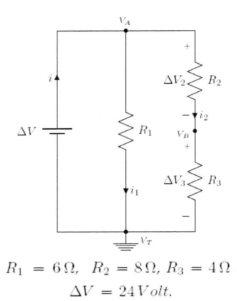

$$R_1 = 6\,\Omega, \quad R_2 = 8\,\Omega, \quad R_3 = 4\,\Omega$$
$$\Delta V = 24\,Volt.$$

Calculate the electric current in each resistor and the potentials of A, B and T.

The resistors with subscripts 2 and 3 are in series therefore:

$$R_{23} = R_2 + R_3 = 12\,\Omega$$

The total resistance of the loop is:

$$\frac{1}{R_{tot}} = \frac{1}{R_1} + \frac{1}{R_{23}} = \frac{1}{4\,\Omega}$$
$$R_{tot} = 4\,\Omega$$

The current flowing out of the generator is:

$$i = \frac{\Delta V}{R_{tot}} = 6\,Ampere$$

Point T has zero potential as it is the earth.
The potential at A will be:

$$V_A = V_T + \Delta V = 24\,Volt$$

The current in the first resistor will be:

$$i_1 = \frac{\Delta V_1}{R_1} = \frac{24\,V}{6\,\Omega} = 4\,Ampere$$

The current in the second resistor will be:

$$i_2 = i - i_1 = 6\,A - 4\,A = 2\,A$$

The potential drop across the second resistor is:

$$\Delta V_2 = R_2 i_2 = 16\,Volt$$

The potential at B is:

$$V_B = V_A - \Delta V_2 = 8\,Volt$$

The potential drop across the third resistor is:

$$\Delta V_3 = R_3 i_2 = 8\,Volt$$

Exercise 20

Given:

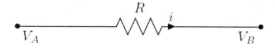

Where R is 6 ohms, the potential at A is 28V and at B is 4V.
We want to feed the circuit for 4 hours.
How much is the energy consumed?
How much charge has passed through the resistance in this time?

The current is given by:

$$i = \frac{\Delta V}{R} = \frac{24\,V}{6\,\Omega} = 4\,A$$

The dissipated power is:

$$P = R \cdot i^2 = 6\,\Omega \cdot 16\,A^2 = 96\,W$$

In 4 hours, the energy consumed is:

$$\Delta E = P \cdot \Delta t = 96\,W \cdot 4\,h = 384\,Wh = 0,384\,kWh$$

The charge that passes through the resistance in 4 hours is:

$$\Delta Q = i \cdot \Delta t = 4\,A \cdot 4\,h = 16\,Ah = 57600\,As = 57600\,C$$

Exercise 21

What is the radius of the circular trajectory of an electron entering perpendicularly into a magnetic field of intensity equal to one millionth of a tesla at a speed of 90,000 meters per second?

The motion of the electron will be circular and therefore the centripetal force will have to balance the Lorentz force generated by the magnetic field.

$$\frac{mV^2}{r} = eVB$$

Recalling the mass and charge values of the electron, we have:

344

$$r = \frac{mV}{eB} = \frac{9.1 \cdot 10^{-31} kg \cdot 9 \cdot 10^4 \frac{m}{s}}{1.6 \cdot 10^{-19} C \cdot 10^{-6} T} = 51.2 \cdot 10^{-2} \, m$$

University-level exercises

Exercise 1

Calculate the electric field and electric potential of an infinite straight wire with linear charge density equal to lambda.

At distance r from the wire we have:

$$E_r = \int_{-\infty}^{+\infty} dz \frac{k\lambda r}{(r^2 + z^2)^{3/2}} = \frac{k\lambda z}{r\sqrt{r^2 + z^2}} \Big|_{-\infty}^{+\infty} = \frac{2k\lambda}{r}, \qquad E_\theta = 0$$

The electric potential for a wire of length 2l is:

$$\varphi(z,r) = \int_{-\ell-z}^{+\ell-z} dz' \frac{k\lambda}{\sqrt{r^2 + z'^2}} = k\lambda \ln\left(z' + \sqrt{z'^2 + r^2}\right)\Big|_{-\ell-z}^{+\ell-z} \simeq \frac{kQ}{\sqrt{r^2 + z^2}}\left(1 + 0\frac{\ell}{d} - \frac{\ell^2}{6}\frac{r^2 - 2z^2}{(r^2 + z^2)^2} + O(\ell^4)\right)$$

Where is it:

$$Q = \lambda 2\ell.$$

Therefore:

$$E_z = -\partial_z \varphi = \frac{k\lambda}{\sqrt{r^2 + (z - \ell)^2}} - \frac{k\lambda}{\sqrt{r^2 + (z + \ell)^2}}, \qquad E_r = -\partial_r \varphi = \dots$$

For an infinite thread we have:

$$\varphi(r) = k\lambda \int_{-\infty}^{+\infty} dz \left[\frac{1}{\sqrt{r^2 + z^2}} - \frac{1}{\sqrt{r_0^2 + z^2}}\right] = -2k\lambda \ln\frac{r}{r_0} = -2k\lambda \ln r + \text{costante}$$

Exercise 2

Calculate the electric field of an infinite plane of radius R with surface charge density equal to sigma.

For reasons of symmetry, the electric field has only a radial part given by:

$$E_r(r) = \frac{1}{4\pi\epsilon_0} \int_0^R 2\pi\rho \, d\rho \frac{\sigma r}{(r^2 + \rho^2)^{3/2}} = -\frac{\sigma}{2\epsilon_0} \frac{r}{\sqrt{r^2 + \rho^2}}\Big|_{\rho=0}^{\rho=R} = \frac{\sigma}{2\epsilon_0}(1 - \frac{r}{\sqrt{r^2 + R^2}}) \overset{R\to\infty}{=} \frac{\sigma}{2\epsilon_0} \operatorname{sgn} r$$

If r is much smaller than R we have:

$$E_r \simeq \frac{Q}{4\pi\epsilon_0 r^2}\left[1 - \frac{3R^2}{4r^2} + \mathcal{O}(R^4)\right]$$

Where is it:

$$Q = \pi R^2 \sigma$$

Exercise 3

Calculate the electric field generated by a uniformly distributed surface charge density on a spherical shell of radius R.

We have:

$$E_r = E_z = \frac{\sigma}{4\pi\epsilon_0} \int_0^\pi \underbrace{R \, d\theta \, 2\pi R \sin\theta}_{ds} \frac{r - R\cos\theta}{[(r - R\cos\theta)^2 + R^2\sin^2\theta]^{3/2}} =$$

$$\begin{cases} 0 & r < R \\ Q/4\pi\epsilon_0 r^2 & r > R \end{cases}$$

Exercise 4

Calculate the potential energy of a sphere of radius R containing a uniformly distributed charge Q.

Inside the sphere the potential is:

$$\varphi(r) = \int E_r dr = \text{cte} - \tfrac{1}{2} q_{in}(r)/(4\pi\epsilon_0 r).$$

Imposing the conditions of continuity for r=R we have:

$$\varphi(r) = \begin{cases} Q(3R^2 - r^2)/8\pi\epsilon_0 R^2 & \text{per } r < R \\ Q/4\pi\epsilon_0 r & \text{per } r > R \end{cases}$$

The potential energy will be:

$$U = \frac{1}{2}\int \rho\varphi\, dV = \frac{1}{4\pi\epsilon_0}\int_0^R \frac{3Q^2 r^2(3R^2 - r^2)}{4R^6} = \frac{3}{5}\frac{Q^2}{4\pi\epsilon_0 R}$$

Exercise 5

Calculate the electric field generated by a sphere of radius R with a surface charge:

$$\sigma(\theta) = \sigma\cos\theta.$$

We can use the principle of superposition of effects.
The given sphere is equivalent to two spheres with opposite charges and uniform densities placed at a distance apart:

$$d\rho = \sigma.$$

A single sphere generates a field within itself:

$$E = r\rho/3\epsilon_0.$$

Two spheres at distance d instead:

$$E = -\rho d/3\epsilon_0 = -P/3\epsilon_0$$
$$P \equiv \rho d$$

Outside the sphere:

$$p = Qd = VP.$$

Where V is the volume of the sphere.
The potential in spherical coordinates will be:

$$\varphi(r,\theta) = \begin{cases} p\cos\theta/4\pi\epsilon_0 r^2 & \text{per } r > R \\ Er\cos\theta & \text{per } r < R \end{cases} = \frac{\sigma\cos\theta}{3\epsilon_0}\begin{cases} R^3/r^2 & \text{per } r > R \\ r & \text{per } r < R \end{cases}$$

Exercise 6

Calculate the force between two electric dipoles at a distance x oriented parallel to their separation.

The electric force between the two dipoles is:

$$F = (p \cdot \nabla)E = p\partial_x E$$

Where the electric field is:

$$E = \frac{1}{4\pi\epsilon_0}\left[\frac{3(p' \cdot r)r}{r^5} - \frac{p'}{r^3}\right] = \frac{1}{2\pi\epsilon_0}\frac{p'}{x^3}\hat{x}$$

So the force along x is:

$$F_x = -3pp'/2\pi\epsilon_0 x^4.$$

Exercise 7

To study a point charge q placed at a distance d from an infinite conducting plane placed at zero potential.

We set the plane at x=0.
Using the image method we have:

$$\varphi(x,y,z) = \begin{cases} \dfrac{q}{4\pi\epsilon_0}\left(\dfrac{1}{|x-d|} - \dfrac{1}{|x+d|}\right) & \text{per } x > 0 \\ 0 & \text{per } x < 0 \end{cases}$$

The surface charge density induced on the plane is:

$$\sigma(y,z) = \epsilon_0 E_\perp = -\epsilon_0 \frac{\partial \varphi}{\partial x} = -\frac{1+1}{4\pi}\frac{dq}{(y^2+z^2+d^2)^{3/2}}$$

From Gauss's theorem, the total charge is:

$$\int \sigma \, dy \, dz = -q,$$

The attractive force on the plane is:

$$F = \int \sigma \frac{E_\perp}{2} \, dx \, dy = \frac{kq^2}{(2d)^2}$$

Exercise 8

Given two plates with surface area S and with defined total charges, calculate the induced electric fields.

Based on how the charges are arranged on the 4 available surfaces, we have the following charge distributions:

$$q - Q, \qquad Q \qquad - Q \qquad q' + Q$$

The electric fields are therefore:

$$E_{\text{sinistra}} = \frac{q - Q}{\epsilon_0 S}, \qquad E_{\text{mezzo}} = \frac{q}{\epsilon_0 S}, \qquad E_{\text{destra}} = \frac{q' + Q}{\epsilon_0 S}$$

The total energy is proportional to the integral of the square of the electric field.

Due to the principle of minimum energy, the charges will tend to minimize this integral, i.e. to arrange themselves symmetrically:

$$Q = (q - q')/2:$$

From which:

$$E_{\text{sinistra}} = E_{\text{destra}} = \frac{q + q'}{2\epsilon_0 S}, \qquad E_{\text{mezzo}} = \frac{q - q'}{2\epsilon_0 S}$$

Exercise 9

An internal conducting wire of diameter d is surrounded by a metal shell of diameter D.

Established the maximum electric field sustainable by the air, find which value of the ratio between the diameters allows to have the maximum potential difference and which the maximum energy.

Using Gauss's theorem, the electric field in the space between the two plates is:

$$E_r = \frac{2k\lambda}{r} = -\frac{\partial \varphi}{\partial r}$$

And so the potential is:

$$\varphi = -2k\lambda\sigma \ln r.$$

The potential difference will be:

$$V = \Delta\varphi = 2k\lambda \ln(D/d)$$

Assuming that the electric field is equal to the maximum:

$$V = E_{\max}\frac{d}{2}\ln\frac{D}{d}$$

So the ratio between the diameters must be:

$$d = D/e.$$

The stored energy is given by:

$$U = \frac{CV^2}{2} = L\pi\epsilon_0 E_{\max}^2 \left(\frac{d}{2}\right)^2 \ln\frac{D}{d}$$

And therefore the ratio between the two diameters must be:

$$d = D/\sqrt{e}.$$

Exercise 10

A cylindrical capacitor of length L and external and internal diameter D ed is maintained at a potential difference V and is immersed in a basin of water of given density.

Calculate how much the water level rises inside the condenser compared to the outside.

If water enters the condenser, its capacity will be:

$$C(z) = 2\pi(z\epsilon + (L - z)\epsilon_0)/\ln(D/d) = C(0) + 2\pi\epsilon_0 z\chi/\ln(D/d).$$

The electric force will therefore be:

$$F_{el} = +\frac{V^2}{2}\frac{dC}{dz} = \frac{\pi V^2 \epsilon_0 \chi}{\ln D/d}.$$

This force will be damped by the gravitational one:

$$F_{grav} = -m(z)g = -\pi\frac{D^2 - d^2}{4}z\rho \cdot g$$

At equilibrium we have:

$$F_{grav} + F_{el} = 0$$

And so the height will be:

$$z = \frac{4V^2\epsilon_0\chi}{(D^2 - d^2)g\rho\ln(D/d)}$$

Exercise 11

A dielectric with a given dielectric constant contains a spherical hole of radius r with a different dielectric constant.

What happens if there is an external electric field?

For the continuity of the electric field at the interfaces it will be necessary that:

$$\epsilon_{out} E^r_{out} = \epsilon_{in} E^r_{in}, \qquad E^\theta_{out} = E^\theta_{in}$$

And then:

$$\epsilon_{out}(E_{ext} - 2\frac{kP}{r^3})\cos\theta = \epsilon_{in} E_{in} \cos\theta, \qquad (E_{ext} + \frac{kP}{r^3})\sin\theta = E_{in}\sin\theta$$

From which:

$$E_{in} = \frac{3E_{ext}}{2 + \epsilon_{in}/\epsilon_{out}}, \qquad \frac{kP}{r^3} = E_{ext}\frac{\epsilon_{out} - \epsilon_{in}}{2\epsilon_{out} + \epsilon_{in}}$$

Which are the internal electric field and the one generated by the dipole.

Exercise 12

Calculate the magnetic field generated by a straight wire of radius a traversed by a constant current density j, surrounded by a cylindrical shell along which an opposite charge flows uniformly.

It is a question of applying Ampere's law to the various cases on the basis of the variation of the radial coordinate.
Inside the thread we have:

$$i = j\pi r^2$$

$$B_r = \mu_0 j r/2 = \mu_0 i r/2\pi a^2.$$

Outside the wire, but for r smaller than the internal radius of the cylindrical shell, we have:

$$i = j \pi a^2.$$

$$2\pi r \, B_r = \mu_0 i$$

Inside the cylindrical shell:

$$B_r = \frac{\mu_0 j}{2\pi r} \pi a^2 \left[1 - \frac{r^2 - b^2}{b'^2 - b^2} \right]$$

Outside the cylindrical shell the magnetic field is zero.

Exercise 13

Calculate the magnetic field generated by a circular coil of radius a in the xy plane through which a current i flows.

Along the axis we have:

$$B_z = \frac{\mu_0}{4\pi} \frac{i}{r^2} 2\pi a \cos\theta = \frac{\mu_0}{2} \frac{ia}{r^2} \cos\theta = \frac{\mu_0}{2} \frac{ia^2}{(a^2 + x^2)^{3/2}}$$

In particular in the center:

$$B = \mu_0 i / 2a.$$

Exercise 14

A charge Q is uniformly distributed on the surface of a homogeneous sphere of mass M and radius R.
The sphere rotates with a given angular velocity.
Calculate the magnetic moment.

An angular ring has radius and area:

$$r = R\sin\theta \; ;$$
$$dS = 2\pi r \cdot R \; d\theta$$

The charge and rotation speed of the ring will be:

$$dQ = Q \; dS/S$$
$$v = \omega r.$$

The current transported will therefore be:

$$di = \frac{dQ \; v}{2\pi r} = \frac{Q\omega}{4\pi}\sin\theta \; d\theta$$

The magnetic moment is obtained by integrating:

$$\mu = \int \pi r^2 \; di = \frac{Q\omega R^2}{4}\int_0^\pi \sin^3\theta \; d\theta = \frac{QR^2}{3}\omega$$

Or:

$$\mu = g\frac{Q}{2M}L$$
$$g = \frac{5}{3}$$
$$L = \frac{2MR^2}{5}\omega$$

Exercise 15

A cylinder of radius R and length I rests on an inclined plane of angle alpha.

355

The cylinder carries a uniform current along its length and there is a vertical magnetic field.
For what current value does the cylinder stand still if friction is negligible?

The magnetic field produces a horizontal force directed towards the inclined plane.

$$dF = \ell B \; di = \ell Bi \; d\theta/2\pi$$

By integrating and projecting the force along the inclined plane, we have:

$$F = \ell Bi \cos \alpha.$$

The component of the gravitational force along the inclined plane is:

$$F = mg \cos \alpha.$$

In equilibrium, these forces are equal.
So the current will be:

$$i = mg \tan \alpha/\ell B.$$

Exercise 16

We want to magnetically accelerate objects along a constant radius.
How should we construct the magnetic field?

The momentum of the object to be accelerated is:

$$p(t) = er B(r, t)$$

And the equation of motion of the object is:

$$\dot{p} = F = eE = e \frac{\dot{\Phi}_B}{2\pi r}$$

Or:

$$\dot{\Phi}_B = 2 \times \pi r^2 \dot{B}.$$

The magnetic field does not have to be constant, but must have a radial profile such that it is greatest at the centre.

Exercise 17

To study the motion of a particle of charge q in a magnetic field:

$$B = (0, 0, B_z(x)).$$

Consider the two half-spaces x>0 and x<0 with magnetic fields equal to:

$$B_1$$
$$B_2 = B_1 + \Delta B$$

The particle moves along semicircles of radius:

$$r_i = mv/qB_i$$

With a frequency:

$$\omega = qB/m$$

Where is it:

$$B \sim (B_1 + B_2)/2$$

With each revolution the particle moves by an amount:

$$\Delta y = 2(r_2 - r_1)$$

That is, it has a drift velocity equal to:

$$v_y^{\text{drift}} \approx \frac{\omega}{2\pi} 2(r_2 - r_1) \approx \frac{v_\perp}{\pi} \frac{B_2 - B_1}{B} = \frac{v_\perp}{\pi} \frac{r}{B} \frac{B_2 - B_1}{r} \approx \frac{m v_\perp^2}{qB} \nabla B$$

Exercise 18

A rectangular circuit of resistance R and sides l fixed and x variable is immersed in an orthogonal magnetic field B.
The length x is varied according to a law x=vt.
Calculate the induced current, the external force, the power required and the power dissipated in the resistor.

From Gauss's theorem the flux is:

$$\Phi = BLvt$$

The induced current will be:

$$I = \mathcal{E}/R = BLv/R.$$

The external force is opposite to the motion and will be:

$$F_{\text{ext}} = BIL = B^2 L^2 v/R$$

The required power is simply, by definition:

$$W_{\text{ext}} = Fv.$$

While that dissipated in the resistance is:

$$W = I^2 R = W_{\text{ext}}.$$

That is, all the work done is dissipated in the resistance.

Exercise 19

Prove that for a magnetic conductor:

$$F = -\tfrac{1}{2}I^2 dL/dx.$$

Where L is the inductance.

The magnetic flux is equal to:

$$\Phi = LI,$$

Magnetic energy is:

$$U = \frac{LI^2}{2} = \frac{\Phi^2}{2L},$$

Inductance is defined as:

$$L = \mu_0 N^2 \frac{S}{d}$$

And the magnetic force is:

$$mbF = +\frac{I^2}{2}\nabla L,$$

By substituting and considering the one-dimensional case we have the thesis.

Exercise 20

A long cylindrical solenoid consists of N turns per unit length wound on an iron core of radius R and length L much greater than R.
Iron has given magnetic permeability and electrical conductivity.
Current is passed through the coils:

$$I = I_0 \cos \omega t.$$

Calculate the magnetic field inside the solenoid, the induced internal electric field and the power dissipated per unit length.

Inside the solenoid we have:

$$B = \mu H = \mu N I$$

While the induced electric field is:

$$\nabla \times E = -\dot{B}$$

Or:

$$E_\theta = \tfrac{1}{2} r \mu n I_0 \omega \sin \omega t$$

This electric field induces a current:

$$J = \sigma E$$

And the energy dissipated equal to JE.

Exercise 21

An area capacitor:

$$S = \pi a^2$$

E distance d much smaller than a, discharges with a given time constant. Calculate the magnetic field and its energy.

The electric charge is:

$$q(t) = q_0 e^{-t/\tau}$$

The electric field is:

$$\epsilon_0 E = \sigma$$

And it generates a uniform displacement current:

$$i_s = \dot{\sigma} = -\dot{q}/S\tau.$$

This current generates a rotating magnetic field:

$$B_\theta = \frac{\pi r^2 \mu_0 j_s}{2\pi r} = \frac{\mu_0 r i}{2S} \qquad r < a$$

Knowing that:

$$i = -S\epsilon_0/\tau,$$

The energy of the magnetic field is:

$$U_B = \int dV \frac{B^2}{2\mu_0} = \frac{d i^2 \mu_0}{16\pi}$$

Exercise 22

361

Given a toroid of rectangular section of dimensions H e (ba) composed of N turns, we have on a cross section, a rectangular turn of given dimensions as in the figure:

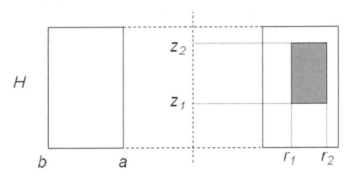

Calculate the mutual inductance between the toroid and the loop if the loop has a constant current.
If instead the coil is crossed by an oscillating current, calculate the electromotive force in the case of oscillation in the z or r direction.

The mutual inductance is calculated taking into account the magnetic field flux generated by the toroid through the coil:

$$M = \frac{\Phi(B_{tor} \mid S_{spira})}{i_{tor}} = \frac{1}{i_{tor}} \iint_{S_{spira}} \vec{B}_{tor} \cdot d\vec{A} = \int_{z_1}^{z_2} \int_{r_1}^{r_2} \frac{\mu_0}{2\pi} \frac{N}{r} dr dz = \frac{\mu_0}{2\pi} N(z_2 - z_1) \log \frac{r_2}{r_1}$$

If the current oscillates in the z direction, there will be no electromotive force as the flow does not vary in that direction.
If instead it is oscillating along r we have:

$$fem = -\frac{d\Phi}{dt} = -\frac{d(Mi_{spira})}{dt} = -\frac{dM}{dt}i_{spira} = -i_{spira}\frac{\mu_0}{2\pi}Nl_z\frac{d}{dt}\left(\log\frac{r_2}{r_1}\right) =$$

$$= -i_{spira}\frac{\mu_0}{2\pi}Nl_z\frac{1}{r_1 r_2}(v_2 r_1 - r_2 v_1) = i_{spira}\frac{\mu_0}{2\pi}Nl_z\frac{v}{r_1 r_2}(r_2 - r_1)$$

Exercise 23

A square coil of side L is mounted on a ribbon which slides with a speed v between the pole pieces of two magnets.

Given l the length of the strip and D the width of the pole pieces, find the value of the electromotive force as a function of time induced in the loop in a time interval equal to a period T of the motion of the strip.

Before the coil enters the pole pieces, the emf is zero.
When the coil is partly inside and partly outside, we have:

$$fem = -\frac{d\Phi}{dt} = -\frac{-BdA}{dt} = \frac{BLvdt}{dt} = BLv$$

When, on the other hand, the coil is totally inside, the emf is still nothing. At the exit we will have a similar situation but with emf of opposite sign. When the loop is out of the pole pieces, the emf is zero again.

Exercise 24

A coil ACD of resistance R has the shape of a circular sector with a central angle at C.
The coil lies on a horizontal plane and can rotate around the fixed vertical axis passing through C.
The loop is partially immersed in a uniform magnetic field B perpendicular to the plane of the loop.
The coil therefore rotates with an instantaneous angular velocity.
Calculate the current induced in the loop when it is completely outside the field B, partly inside, partly outside or completely inside the field.

The induced current is proportional to the electromotive force.
Since the latter is null when the loop is totally outside or totally inside, in these cases there is no induced current (since the flux of the magnetic field does not vary).
If the loop is partially in the field, we have:

$$fem = -\frac{d\Phi}{dt} = -\frac{BdA}{dt} = -\frac{Ba^2 d\theta}{2dt} = -\frac{1}{2}Ba^2\omega.$$

And then:

$$i = \frac{fem}{R} = -\frac{Ba^2}{2R}\omega.$$

Exercise 25

Given a square coil with side L and resistance R and a wire carrying a current in the z direction.

Given a and b the distance of the nearest and farthest parallel side of the coil with respect to the wire, move the coil with speed v towards the wire.

Find the flux of the B field through the loop, the emf and the current induced in the loop.

Finally, find the force acting on the coil.

The magnetic field is given by the Biot-Savart law, the flux is therefore:

$$\Phi(B) = \int_{spira} \vec{B} \cdot d\vec{a} = \int_a^b \frac{\mu_0}{2\pi} \frac{i}{r} L dr = \frac{\mu_0}{2\pi} iL \log\frac{b(t)}{a(t)}$$

The electromotive force is simply the derivative of this flux:

$$fem = -\frac{d\Phi(B)}{dt} = -\frac{\mu_0}{2\pi} iL \frac{a(t)}{b(t)} \frac{b'(t)a(t)-b(t)a'(t)}{a^2(t)} = \frac{\mu_0}{2\pi}iLv\frac{b(t)-a(t)}{a(t)b(t)} = \frac{\mu_0}{2\pi}\frac{iL^2v}{a(t)b(t)}$$

The induced current is:

$$I = \frac{fem}{R}$$

Parallel forces act in opposite directions on the loop:

$$F_b = ILB(b(t)) \qquad\qquad F_a = ILB(a(t))$$

And so the total force is:

$$F_{tot} = F_a - F_b = IL\left[B(a(t))-B(b(t))\right] = IL\frac{\mu_0}{2\pi}i\left|\frac{1}{a(t)} - \frac{1}{b(t)}\right| = \frac{\mu_0}{2\pi}\frac{iIL^2}{a(t)[a(t)+L]}$$

Exercise 26

A circular loop of radius a and resistance R is thrown towards a magnetic dipole of value m with a given initial velocity.
The axis of the dipole and the coil coincide.
Calculate the flux, the induced current and the force acting on the loop.

By placing:

$$K = \frac{\mu_0}{4\pi} m$$

And defining z the distance between the center of the loop and that of the dipole and r the distance from the z axis, the expression of the dipole field in cylindrical coordinates is:

$$\vec{B} = \begin{pmatrix} B_r \\ B_\varphi \\ B_z \end{pmatrix} = \frac{K}{\left(z^2 + r^2\right)^{3/2}} \begin{pmatrix} \dfrac{3rz}{z^2 + r^2} \\ 0 \\ \dfrac{3z^2}{z^2 + r^2} - 1 \end{pmatrix}$$

From which, neglecting the quadratic terms, we have:

$$\vec{B} \approx \frac{K}{z^3} \begin{pmatrix} \dfrac{3r}{z} \\ 0 \\ 2 \end{pmatrix}$$

The flow acting on the coil is:

$$\Phi(B) = \iint\limits_{C(spira)} \vec{B} \cdot d\vec{a} = \iint\limits_{C(spira)} B_z da = 2\frac{K}{z^3} \iint\limits_{C(spira)} da = 2\frac{K}{z^3} \pi a^2$$

The induced current is:

$$i = -\frac{1}{R}\frac{d\Phi(B)}{dt} = -\frac{1}{R}\frac{d}{dt}\left(2\frac{K}{z^3}\pi a^2\right) = -\frac{1}{R}2K\pi a^2\frac{d}{dt}\left(\frac{1}{z^3}\right) =$$

$$= \frac{6K\pi a^2}{R}\frac{1}{z^4}\frac{d}{dt}(z) = \frac{6K\pi a^2}{Rz^4}v$$

The force acting on the loop is:

$$\vec{F} = \oint_{spira} i\,d\vec{l} \times \vec{B}$$

Decomposing along the axes:

$$\hat{r}F_r + \hat{k}F_z = \oint_{spira} i\,d\vec{l} \times \vec{B}_z + \oint_{spira} i\,d\vec{l} \times \vec{B}_r$$

From which:

$$F = F_z = -\oint_{spira} i\,dl B_r = -i\frac{3Ka}{z^4}\oint_{spira} dl = -\frac{3Ka}{z^4}i2\pi a = -\frac{6K\pi a^2}{z^4}i = -\left(\frac{6K\pi a^2}{z^4}\right)^2\frac{v}{R} > 0$$

Exercise 27

Given a circular loop of radius a, current I and angular momentum m, at a sufficiently large distance the magnetic field is given by:

$$\vec{B}_1 = \begin{pmatrix} B_r \\ B_\varphi \\ B_z \end{pmatrix} = \frac{\mu_0}{4\pi}m\begin{pmatrix} \dfrac{3r}{z} \\ 0 \\ 2 \end{pmatrix}\frac{1}{z^3}$$

At a distance D from this coil another coil of radius b and given resistance is placed.
If the current induced in the second turn is given by:

$$I_2(t) = I_{2,0} \sin \omega t$$

Find the current in the first coil.

From Faraday's law we have that:

$$I_2 = \frac{fem_2}{R_2} = -\frac{1}{R_2}\frac{d\Phi}{dt} = -\frac{1}{R_2}\frac{d}{dt}\iint_{S_1} \vec{B}_1 \cdot d\vec{a}_2 =$$

$$= -\frac{1}{R_2}\frac{d}{dt}\iint_{S_1} B_z r dr d\varphi = -\frac{1}{R_2}\frac{d}{dt} 2\pi \int_0^b \frac{\mu_0}{4\pi} m \frac{2}{D^3} r dr =$$

$$= -\frac{1}{R_2}\frac{\mu_0}{2\pi D^3}\frac{d}{dt}\left(\pi a^2 I_1 \pi b^2\right) = -\frac{1}{R_2}\frac{\mu_0}{2D^3}\pi a^2 b^2 \frac{dI_1}{dt} = -K \frac{dI_1}{dt}$$

Therefore, integrating, we find:

$$I_1(t) = -\frac{1}{K}\int I_2(t) dt + const. = -\frac{1}{K}\int I_{2,0} \sin \omega t dt + const. =$$

$$= \frac{\omega}{K} I_{2,0} \cos \omega t + const.$$

Exercise 28

Write explicitly the electric field and magnetic field for a plane monochromatic wave traveling in vacuum in the direction of the negative x axis and polarized in the z direction.

In general we have:

$$\begin{aligned}
\mathbf{E}(\mathbf{r}, t) &= \hat{\mathbf{n}}\, E_0 \cos(\mathbf{k} \cdot \mathbf{r} - \omega t) \\
\mathbf{B}(\mathbf{r}, t) &= (\hat{\mathbf{k}} \times \hat{\mathbf{n}})\, \frac{1}{c} E_0 \cos(\mathbf{k} \cdot \mathbf{r} - \omega t) \\
|\mathbf{k}| &= \frac{\omega}{c}\ ;
\end{aligned}$$

Considering that:

$$\hat{\mathbf{k}} = (-1, 0, 0),\ \hat{\mathbf{n}} = (0, 0, 1).$$

From which:

$$(\hat{\mathbf{k}} \times \hat{\mathbf{n}}) = \hat{\mathbf{y}}$$

You get:

$$\begin{aligned}
\mathbf{E}(\mathbf{r}, t) &= \hat{z}\, E_0 \cos(-kx - \omega t) = \hat{z}\, E_0 \cos(kx + \omega t) \\
\mathbf{B}(\mathbf{r}, t) &= \hat{y}\, \frac{1}{c} E_0 \cos(kx + \omega t)\ .
\end{aligned}$$

Exercise 29

For a monochromatic plane wave traveling in the z-direction linearly polarized in the x-direction, find all elements of the Maxwell stress tensor:

$$T_{\alpha\beta} = \begin{pmatrix} T_{xx} & T_{xy} & T_{xz} \\ T_{yx} & T_{yy} & T_{yz} \\ T_{zx} & T_{zy} & T_{zz} \end{pmatrix}$$

In vacuum we have:

$$T_{\alpha\beta} = \epsilon_0 \left(E_\alpha E_\beta - \frac{1}{2}\delta_{\alpha\beta}\mathbf{E}^2 \right) + \frac{1}{\mu_0} \left(B_\alpha B_\beta - \frac{1}{2}\delta_{\alpha\beta}\mathbf{B}^2 \right)$$

$$\mathbf{E}(\mathbf{r},t) = \hat{\mathbf{x}}\, E_0 \cos(kz - \omega t)$$
$$\mathbf{B}(\mathbf{r},t) = \hat{\mathbf{y}}\frac{E_0}{c} \cos(kz - \omega t) \ ,$$

Where in our case:

$$T_{xx} = \epsilon_0 \left(1 - \frac{1}{2} \right) E_0^2 \cos^2(kz - \omega t) + \frac{1}{\mu_0} \left(0 - \frac{1}{2} \right) \frac{E_0^2}{c^2} \cos^2(kz - \omega t) =$$

$$= \frac{1}{2} \left(\epsilon_0 - \epsilon_0\mu_0 \frac{1}{\mu_0} \right) E_0^2 \cos^2(kz - \omega t) = 0 \ ; \quad T_{xy} = 0 \ ; \quad T_{xz} = 0 \ ;$$

$$T_{yx} = 0 \ ; \quad T_{yy} = \frac{1}{2} \left(-\epsilon_0 + \epsilon_0\mu_0 \frac{1}{\mu_0} \right) E_0^2 \cos^2(kz - \omega t) = 0 \ ; \quad T_{yz} = 0 \ ;$$

$$T_{zx} = 0 \quad T_{zy} = 0 \ ; \quad T_{zz} = \frac{1}{2} \left(-\epsilon_0 - \epsilon_0\mu_0 \frac{1}{\mu_0} \right) E_0^2 \cos^2(kz - \omega t) \ ;$$

From which:

$$T_{\alpha\beta} = \begin{pmatrix} 0 & 0 & 0 \\ 0 & 0 & 0 \\ 0 & 0 & -\epsilon_0 E_0^2 \cos^2(kz - \omega t) \end{pmatrix} .$$

12

CRISIS OF CLASSICAL PHYSICS

Introduction

In the second half of the 19th century, it became evident that classical physics itself had very large problems in explaining physical reality.

After two centuries of speculations and theories that had ranged from one end of scientific knowledge to the other, a point of no return had been reached in which these problems gradually became more and more in consequences.

There were various scientific fields that brought a series of experiments and data in contrast with the classical system, among which we recall astronomy, chemistry and many parts of physics, such as electromagnetism.

On closer inspection, there were at least six different reasons that troubled scientists and that could lead to the prediction of new, more complex and more general theories.

Let's list these six reasons.

Astronomical observations

The Newtonian theory of gravitation was no longer in line with some experimental verifications of an astronomical nature, mainly with a specific motion of the planets, in particular Mercury, called the precession of perihelia.

Electromagnetism and invariant transformations

The second problem of classical physics was instead typically linked to electromagnetism, above all when its reconciliation with Newtonian mechanics was attempted.

It seemed quite natural to extend to electromagnetism the concepts of Galilean relativity that suited Newtonian mechanics so well.
In particular, as we have seen in the first paragraph, the transformations which guarantee the transition from one inertial frame to another are

$$t' = t$$

$$x' = x - vt$$

where v was the relative velocity which was compounded with the well-known parallelogram rule. On the basis of these transformations, Newton had stated that space and time were absolute concepts and, until then, there had been no denial of this.
Surprisingly, however, Maxwell's equations were not invariant with respect to Galilean transformations and therefore something had to be revised.
In mathematical terms, the solutions of Maxwell's equations were not part of the Galilean group.

In 1900, Lorentz discovered that Maxwell's equations were invariant with respect to other transformations, those which we define as Lorentz transformations and which are given by (limiting ourselves to a single spatial dimension):

$$t' = \frac{1}{\sqrt{1 - \frac{v^2}{c^2}}} \left(t - \frac{v}{c^2} x\right)$$

$$x' = \frac{1}{\sqrt{1 - \frac{v^2}{c^2}}} (x - vt)$$

In reality those transformations were already discovered in 1887 by Voigt and reworked by Larmor in 1897, but it was Lorentz who linked them to the invariance of Maxwell's equations and it was only Poincaré 1904 awho gave an easy-to-use formulation, which is the one we still use today.
We immediately point out that at low speeds, ie not comparable with those of light, the Lorentz transformations are reduced to the Galilean ones.
At this juncture it should have been understood that they represented an extension of it and therefore a basis for a much broader theory which included Newton's mechanics as a limiting case.

But the times, and the brains, were not yet ripe for this epochal leap.

Since the publication of Maxwell's equations in 1864, there was a very pressing question in the world scientific community: should Galilean relativity and with it Newtonian mechanics be thrown overboard or should Maxwell's equations be admitted to be groundless?

Nobody took such an extreme position because it was a question of going against the experimental data. Both theories explained many effects and perfectly predicted many empirical results.

The easiest way was to find a shortcut to reconcile the two theories with a single explanation, also because there was always the problem of the instantaneous action of the gravitational force at a distance.

Precisely for that reason, the existence of a " *luminiferous ether* " was hypothesized which should have been the solution to everything: the means by which electromagnetic waves were transmitted (since one still had an ancestral fear of the void, a *horror vacui* of philosophical reminiscences) and also squaring the circle to save Galilean relativity with its transformations in terms of space and time and Newton's mechanics with the law of universal gravitation.

However, in 1887 the experiment of Michelson and Morley placed a tombstone on the presumed existence of the ether.

In an ingenious way, an optical system had been set up which first combined the separation of two light rays which would have followed two distinct paths (one perpendicular to the other) and then recombined them, always optically, and studied the interference figure as in a normal interferometer .

Knowing the physical length of each of the two paths and measuring the time that the light rays would have taken to make the two different paths, that experiment aimed to unravel the last doubts about the existence of the ether and the composition of speeds.

It was shown that the simplest explanation ever to clarify the physical result obtained was to assume the constancy of the speed of light in all directions, independently of the motion conditions of the observer and of the reference system.

Light therefore, like any electromagnetic wave, does not satisfy the transformations of Galilean relativity.

The question therefore remained open above all because the electromagnetic waves respected the speed composition rule as Lorentz had foreseen and as had been verified by the Michelson and Morley experiment and not as stated by Newtonian mechanics and Galilean relativity.

The spectrum of the black body

The experimental verifications on the black body spectrum were one of the salient points for the undermining of the previous theory.

A black body is an ideal object that absorbs all the amount of electromagnetic radiation and does not transmit or reflect any kind of energy.

This is obviously a theoretical abstraction, first studied by Kirchhoff in 1862, but in nature there are physical objects that come close to this definition, for example a hollow object kept at a constant temperature.

The problem arose from the fact that, using Maxwell's equations, the experimental data did not coincide with what was expected, especially at low wavelengths (this is known as the ultraviolet catastrophe).

The theory correctly predicted that the intensity of black body radiation depended only on the temperature of the same and aligned with the experimental data for long wavelengths (from infrared on), but predicted an infinite divergence of this intensity for low wavelengths, when reality said it should go to zero.

So something didn't add up at all and we had to go further.

Photoelectric effect

The photoelectric effect was discovered by Hertz in 1887 while he was studying the propagation of electromagnetic waves in air, intent on verifying Maxwell's equations.

Taking two electrically charged conductive bodies it was seen that the electric discharges between them were much more intense if these bodies were exposed to ultraviolet radiation.

The surprising discovery was that these discharges disappeared completely under a certain radiation threshold, regardless of the intensity of the same.

This meant that it was not the power of the electromagnetic wave that was decisive, but its wavelength or frequency, quantities that are completely assimilable since the fundamental relationship of the waves always exists:

$$\lambda f = v$$

where v is the speed of wave propagation.

In the case of electromagnetic ones it is equivalent to the speed of light which is a constant.

Furthermore, the energy of the electrons emitted by these discharges also depended proportionally on the frequency which, therefore, played a primary role both in deciding when it was possible to emit those discharges and in determining the emission energy itself.

All this could not be explained through Maxwell's equations nor using classical thermodynamics, deriving from Newtonian mechanics.

Stability of matter

A further problem was given by the evident stability of the atom and of matter in spite of what electromagnetism foresaw.

The discovery of atoms in the modern era dates back to 1803 by Dalton, while throughout the nineteenth century he worked on the establishment of the periodic table of elements, discovering all those characteristics of repetition, precisely of periodicity, between contiguous elements or elements belonging to the same group , without however being able to give a convincing theoretical explanation.

Furthermore, the famous octet rule, which structures and defines precisely the periodicity of the table, was not at all clear and was mainly used for its extraordinary empirical validity.

But the real problem was given by the structure of the atom, the constituent basis of all molecules and all matter.

With the discovery of the electron in 1874 by Stoney and the subsequent discoveries of the nucleus of the atom by Rutherford and the electric charges associated with it (positive for the nucleus and negative for the electron), a fundamental question arose .

Assuming that electrons (negatively charged) orbit the nucleus (positively charged) in the same way as planets orbit the Sun, the Lorentz force in such an electromagnetic field led to the startling conclusion that the atom would have to collapse in very few instants (fractions of a second), in stark contrast to the stability of matter.

So what was wrong with this reasoning? Maybe electromagnetism needed to be reviewed?

As we will see in the next paragraph, and almost surprisingly, the same theory that was able to concretely explain the black body spectrum would also solve this problem.

The wave-particle dualism

The last inconsistency we present here appears to be, at first blush, a more philosophical than practical issue.

In science there are problems that transcend the mere experimental aspect and that reappear in the form of logical questions.

However, the importance of these questions is no less than the answers that emerge from the experiments, given that it is precisely from these contradictions of a theoretical nature that ideas are drawn for overcoming the previous theories.

Well before Maxwell's equations, everyone knew the atavistic problem of the nature of light, whether it was wave or corpuscular.

The discoveries regarding optics and light about the properties of interference, reflection, refraction and diffraction led to a totally wave-like nature of this branch of physics.

On the other hand, matter understood as an aggregate of molecules, atoms and electrons was, without a shadow of a doubt, of a more corpuscular nature.

But Maxwell's equations shuffled the cards.

Those equations, referring to the electromagnetic field, explained very well physical phenomena such as electricity and magnetism which, however, at first glance, had two different natures.

All electrical phenomena were in some way similar to corpuscles, the electric current itself was nothing but a mass of moving electrons, while the magnetic ones referred to wave theories.

Furthermore, the solutions of these equations are electromagnetic waves that propagate with a speed equal to that of light, and the theory of the electromagnetic field encompasses optics as a particular sector of physics.

Hence the considerable doubt and a sort of dualism between corpuscular and wave nature, aggravated by the just mentioned photoelectric effect, in which the electromagnetic radiation (and therefore also light) seems to behave like a particle, therefore a material object, while in many other experimental evidence as a wave, therefore as something that does not transport matter but only energy (from the well-known definition of wave). Even this dualism will be resolved by the theory that we will shortly explain, in particular by relating a typically wave-like quantity such as wavelength or frequency with one linked to the material world, such as momentum.

Towards the new

These inconsistencies go under the generic name of "crisis of classical physics" and this happened at the end of the 19th century.

It was the new century, the Twentieth, that opened with two theories destined to upset the entire previous system.

At the beginning of the twentieth century, therefore, there was no single structured physical theory capable of explaining these problems.

This observation prompted many scientists to investigate the nature of things again and to inaugurate one of the most profitable seasons in the history of physics, an extraordinary thirty years that began in 1900.

Lightning Source UK Ltd.
Milton Keynes UK
UKHW020936090123
415051UK00017B/1172

9 798215 134085